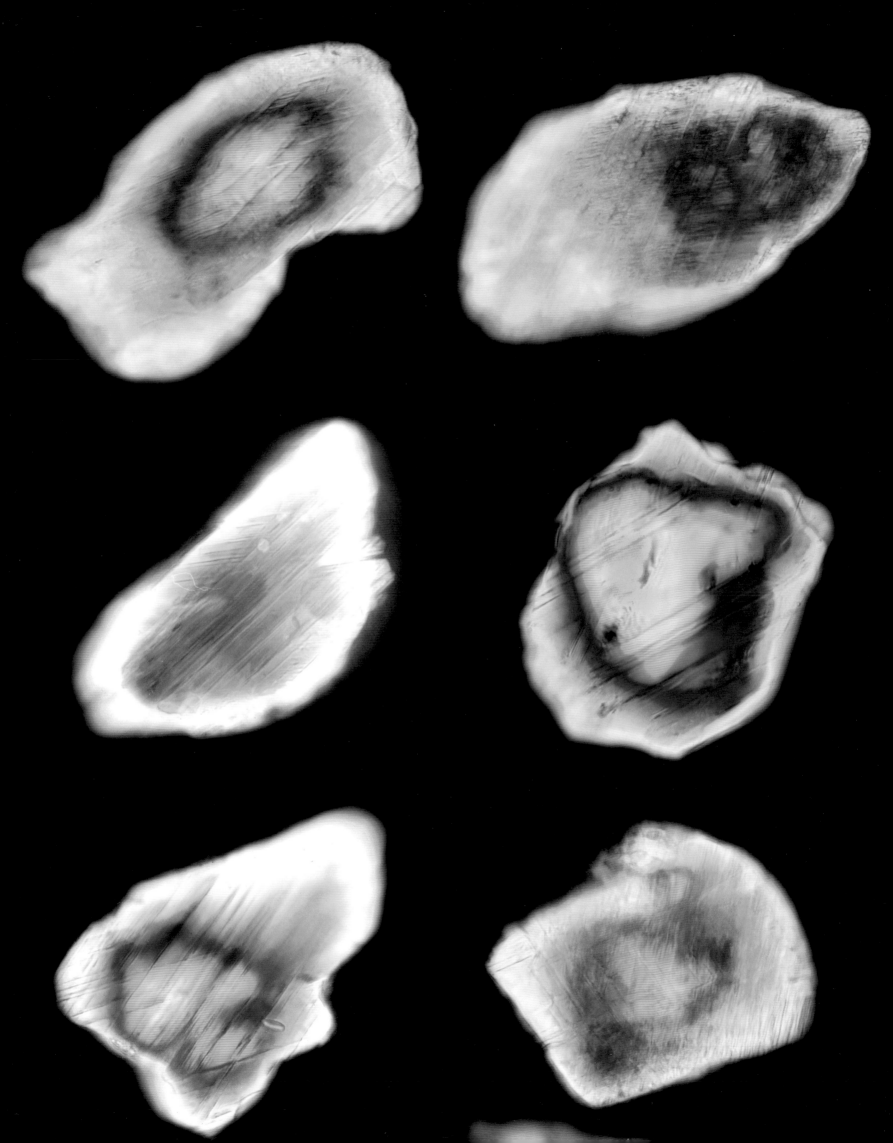

A VOYAGE THROUGH SCALES
The Earth System in Space and Time

T

T

L

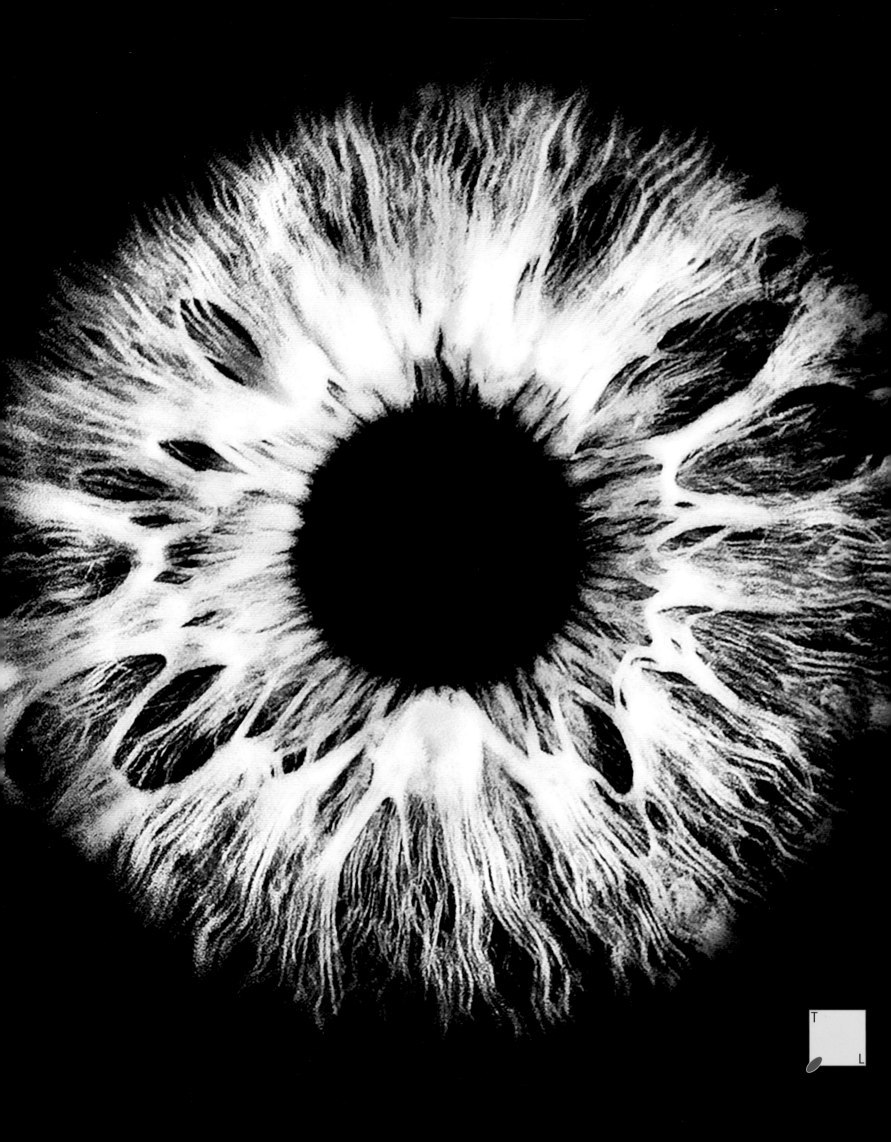

Patterns of billions of stars on the night skies, cloud patterns, sea ice whirling in the ocean, rivers meandering in the landscape, vegetation patterns on hillslopes, minerals glittering in the sun, and the remains of miniature creatures in rocks — they all reveal themselves as complex patterns from the scale of the universe down to the molecular level. A voyage through space scales.

From a molten Earth to a solid crust, the evolution and extinction of species, climate fluctuations, continents moving around, the growth and decay of ice sheets, the water cycle wearing down mountain ranges, volcanoes exploding, forest fires, avalanches, sudden chemical reactions — constant change taking place over billions of years down to milliseconds. A voyage through time scales.

A Voyage Through Scales is an invitation to contemplate the Earth's extraordinary variability across space and time scales. Depending on the view point, different features are revealed. Big structures and small structures, elongated in shape, connected, regular sinusoidal forms, vortices, branching structures or the more geometric shapes imposed by humans. They are a legacy of the intertwined processes of the Earth on its voyage through time. We can use them to understand how we arrived at what we see now, or we can use them to predict where we will go in the future.

This book presents photographs from around the world. Each photo comes with a yellow inset diagram with a small blue ellipse indicating the space and time scales of the processes depicted on the photo. The space scales relate, for example, to the diameter of a vortex, the time scales to its life time. The ranges of the diagrams were selected to conform to what humans can experience directly - from a millimetre to the diameter of the Earth, and from seconds to our immediate history of a few hundreds of years. Where measurement instruments are shown, the diagrams give the sampling scales as small squares. The introductory chapter provides a more detailed description of the notion of space and time scales.

This material is presented through the prism of the journals of the European Geosciences Union. The main body of the book is organized into fourteen chapters. Each addresses space-time scales in the field of one of the journals, from Annales Geophysicae that deals with the Sun-Earth system to The Cryosphere that is dedicated to all aspects of frozen water and ground. The authors were asked to interpret the scale issues in their respective fields that challenge our ability to measure, to model, to comprehend.

The final chapter highlights the fundamental editorial concept of these journals — Open Access. The European Geosciences Union, in cooperation with Copernicus Publications, has been a pioneer of Open Access Publication, having published open access journals since 2001, a concept that, since then, has spread to most fields of science.

The editors would like to thank the many people who contributed to bringing this volume to reality, in particular the Editors of the journals and other authors for their insightful pieces, Lois Lammerhuber for designing this book, Shaun Lovejoy for suggesting the motto, the production staff for their professional approach, and — last but not least — Thomas Hofmann who conceived the original idea of the book and coordinated the entire project.

As the Earth and humankind continue their voyage through time we have arrived at a stage where the human imprint on the Earth system can no longer be overlooked. It may no longer suffice to treat humans as boundary conditions in an isolated way but as an integral part of the coupled human-nature system. As this coupling is becoming more and more central to Earth science, so the coupling between the geoscience disciplines becomes more important. Perhaps the concepts of scale can play a catalytic role in our endeavour of integrating our disciplines into Earth system science to better understand the interplay of processes across scales.

Günter Blöschl, Hans Thybo, Hubert Savenije

Cover Circumantartic circulation
Endpaper 1 Milky Way above the Danube River
Endpaper 2 Shocked quartz crystals, Bosumtwi-Crater
2 Human Iris

A VOYAGE THROUGH SCALES
The Earth System in Space and Time

Günter Blöschl Hans Thybo Hubert Savenije

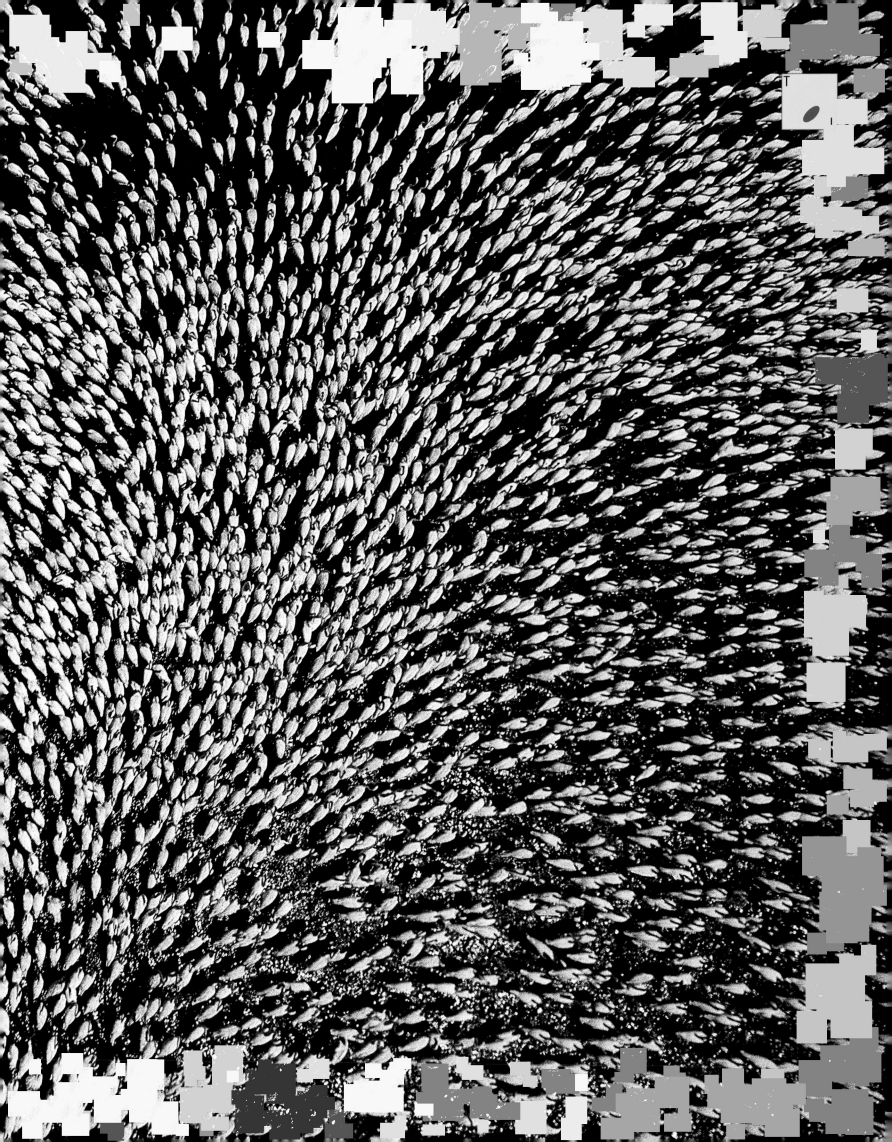

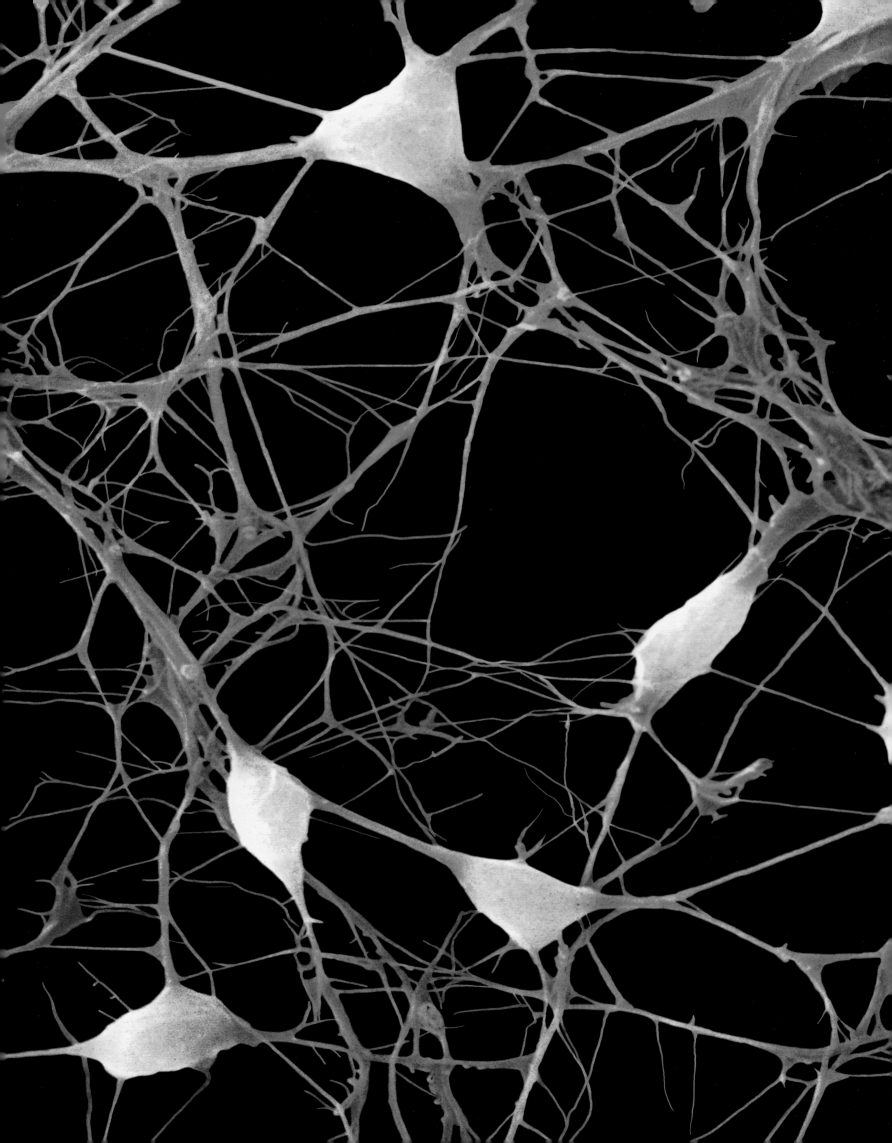

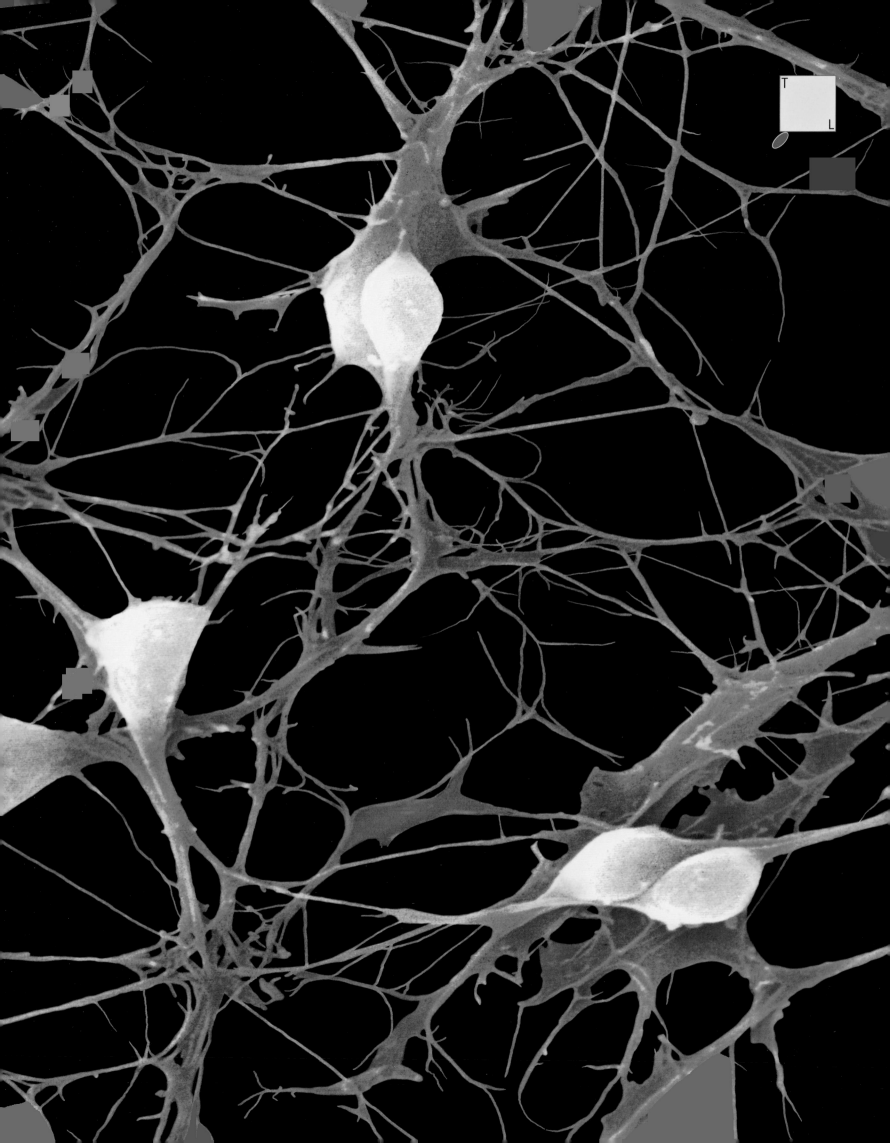

TABLE OF CONTENTS

A VOYAGE

Günter Blöschl, Hans Thybo, Hubert Savenije, Shaun Lovejoy

The evolution of the Earth, the planetary system, the entire universe and indeed humankind itself is a voyage. A voyage through space and time. Ever since the formation of the universe some 14 billion years ago and the formation of the Earth some five billion years ago there has been constant change. The young molten Earth slowly cooled with time to form a solid crust that allowed liquid water to exist on its surface. When photosynthetic life appeared some two billion years ago the atmosphere became enriched with oxygen. Life responded to the ever-changing planet by continuously evolving. Plate tectonics has shaped the face of the Earth, its oceans and continents, and subsequently, the life that dwells in them. In turn, the biosphere has had a major effect on the atmosphere and other conditions on Earth, such as the formation of the ozone layer, the nature of the air we breathe, and the creation of soil.

The interplay of these processes across the Earth system continues to shape the Earth as we know it today. The Earth's lithosphere is broken up into plates that float upon the much thicker layer of viscous material that makes up most of the Earth's mantle. Plates move at speeds of a few centimetres per year, comparable to the growth rate of human hair and nails. Convection in the mantle is believed to play a role in driving plate tectonics, although the main forces appear to originate in the plates themselves. Metamorphic reactions in subducting plates increase density, pulling the plates down, and the high topography at mid-oceanic ridges lead to outward forces that push the plates away. Most earthquakes are concentrated along plate boundaries. Mountain building and volcanism are all closely associated with these processes. Unseen to the eye, the Earth's magnetic field protects humans from the solar wind that would otherwise strip away the upper atmosphere, including the ozone layer that protects the Earth from harmful ultraviolet radiation. The geomagnetic field and its polarity reversals are influenced by electrical currents in the solid inner and fluid outer core. Higher up in the atmosphere, radiation and convection regulate the temperature at the Earth's surface, controlling the presence of snow and ice cover which, in turn, cools the Earth due to its low reflectivity. Wind patterns regulate oceanic upwelling that brings nutrients to the marine biosphere and determine the spatial distribution of rain that sustains the terrestrial biosphere. Rain, frost, and wind erode the Earth's crust, reshaping the land surface, and replenishing the soils with the supply of elements needed to sustain life. The soils are the life support system of the Earth. Home to an entire universe of microorganisms they help make our air breathable, clean the water we drink and support production of the food we eat. As the waters in the landscape converge, rivers are formed — the arteries of life. Most ancient civilisations have developed along rivers, and rivers still play a key role in today's society be it through water supply or transport but, in turn, threaten civilisation through floods. Rivers close the water cycle from the atmosphere to the ocean and continuously wash sediments into the ocean. The oceans themselves play a pivotal role in the Earth system due to their large storage capacity of heat, carbon and other quantities, their exchange of heat between hot and cold regions, their supply of moisture to the continents, and their position at the interface between the spheres.

The rich diversity of these interlinked processes manifests itself in patterns. The amazing patterns we see when we look around us, patterns that this book attempts to capture. Patterns of billions of stars and untold galaxies in the night skies, cloud patterns, and perhaps the spectacular view of an aurora reflecting the interactions of the solar wind with the magnetic field of the Earth. Satellite images may show patterns of a threatening cyclone waiting to wreak havoc across the land. When on an aeroplane flight we may see amazing patterns of nightlights reflecting the preferences of modern human civilisation, the patterns of mountain ridges and valleys and, in the ocean, the tortuous paths of algae or sea ice whirling along complex currents. During a leisurely weekend stroll we may admire from a mountain summit the rivers meandering in the landscape, the vegetation patterns before us that show the clear mark of humans, and diverse soils of stunningly different colours. In front of us the mountain faces with their undulating patterns of layers and faults. Closer up, minerals may glitter in the sun exhibiting intricate crystal structures, and the remains of corals and other miniature creatures our rocks are composed of reveal themselves as complex patterns. Aided by microscopes and modern instrumentation we penetrate deeper into a genuine micro-cosmos of patterns of life revealing themselves to the observer down to the molecular level.

This is the voyage through space, through space scales, from the universe to molecules. A voyage of patterns viewed from far away down to patterns viewed up close. Depending from where we look, we see different features. Big structures and small structures, elongated in shape, connected, regular sinusoidal forms, vortices, branching structures or the more geometric shapes imposed by humans. Sometimes the patterns are similar to an enlarged part of itself (i. e. the whole has the same shape as one of the parts), a notion termed self-similarity. More frequently, such scale invariance is only established if the enlargement is squeezed and rotated. This is the case for coastlines and clouds, for example. Whatever the patterns look like, they are a reflection of processes occurring at different scales — or sometimes — a single dynamical mechanism that operates over a wide range of scales. Wherever two media interact under some driving gradient (water and air, air and land, water and sediments, continental shelves, and even traffic over a road layer), patterns appear. The patterns we observe are a legacy of the voyage through the life time of the Earth. We can use them to understand how we arrived at what we see now, or we can use them to predict where we will go in the future. For every pattern we can see, there are many more patterns we cannot see. How do we go from what we can see to what we cannot see? How do we use these patterns?

There are many ways of using these patterns in understanding the Earth system, to test our theories and to infer the dynamics of the pattern forming processes. One particularly interesting attribute are the *scales* associated with these patterns. How big are the features we see in the patterns? How long do the processes take? Or in other words, what are the space scales and time scales of the processes and what is the relation between the two? These scales can be quantified so we can plot them on graphs. For example, we may measure the diameter of a vortex, the length of a dune, the width of a stream — their space (or length) scales. And we can measure the lifetime of the same vortex, and how long it takes to form the dune or the stream — their time scales. Time scales may also relate to the periods of periodic phenomena such as in tides, and they may relate to fluxes. For example, the cycling of water in the Earth system occurs at different rates and water molecules may stay for a longer time in some parts of the system than in others. The residence times of water in the atmosphere are days, in rivers weeks, in aquifers hundreds of years or more, in the oceans thousands of years, in the ice sheets tens of thousands of years, and in the crust and mantle water stays even longer.

These space and time scales are extremely useful concepts when looking at the Earth system in its entirety. As opposed to the focus on a single process and scale, they open up a whole panorama of different processes that operate at different scales and interact across them. The energy cascade of turbulence is an example of such interactions where the solar energy drives circulations at the global scale which in turn drives consecutively smaller scale motions of the atmosphere and the ocean down to little whirls. As the cascade proceeds to smaller and smaller scales, the variability systematically builds up and at millimetre scales, it can be enormous. Other quantities can cascade as well through scale interactions. For example, when stream sediments are broken into consecutively smaller pebbles through the flow of water, mass cascades towards smaller sizes. When water droplets in clouds coalesce, mass cascades towards larger scales.

Henry Stommel was probably the first Geoscientist to plot the time scales of geo-processes against their space scales in his 1963 Science paper on Varieties of Oceanographic Experience. On p. 572 he noted, "It is convenient to depict these different components of the spectral distribution of sea levels on a diagram ([his] Fig. 1) in which the abscissa is the logarithm of period, P in seconds, and the ordinate is the logarithm of horizontal scale, L in centimetres. If we knew enough we could plot the spectral distribution quantitatively by contours on this period-wavelength plane."

In this book, the "Stommel diagram" is used to visualize the space and time scales of the processes leading to the patterns shown in the photos. The scales are orders of magnitudes and, as more than one process may be associated with a particular pattern, they may not always be unique. Still, they are indications of the powers of ten scales. In line with Stommel's idea, each photo has a small inset diagram, with length scale L on the horizontal axis, time scales T on the vertical axes and the small blue ellipse indicates the particular space-time scale combination. In most instances the ellipses have been plotted in an oblique position to indicate that larger space scales are associated with larger time scales. Some phenomena — notably turbulent ones — line up quite linearly on Stommel diagrams so that their size-duration relationships are power laws reflecting the same underlying dynamical mechanism over a range of scales, and there is one diagram (on p. 18) where the scales have been plotted as a blue line. In some figures the ellipse has been plotted horizontally to indicate that the time scale does not increase with space scale; these relate to diurnal or annual variations of solar radiation. For clarity, all ellipses have the same size.

The yellow-blue diagrams below show the ranges chosen for the small diagrams throughout the book: Length scales from 1 mm to 10 000 km, and time scales from 1 sec to 300 years, with a centre point of 100 m and 1 day. These ranges were selected to conform to what humans can experience directly. With the naked eye we can discern things in size from a millimetre to the diameter of the Earth, and we can discern seconds to our immediate history of a few hundreds of years. If the little ellipse on the left plots within the diagram, then the photo is within the realm of humans' direct comprehension, if it plots outside, instrumentation needs to be used to extend our sensory abilities.

The ratio of the space and time scales of a process gives its characteristic velocity, as indicated by the oblique lines in the diagram below. For, say, regional atmospheric processes it is on the order of 10 m/s, for river flow it is around 1 m/s, for subsurface flow it is 100 m/day to 1 cm/day. The movement of continental plates is even slower with a few centimetres per year and the landscape evolves with rates of millimetres per century. The characteristic velocities may relate to the speed of particles, the speed by which waves propagate or the rate by which structures grow.

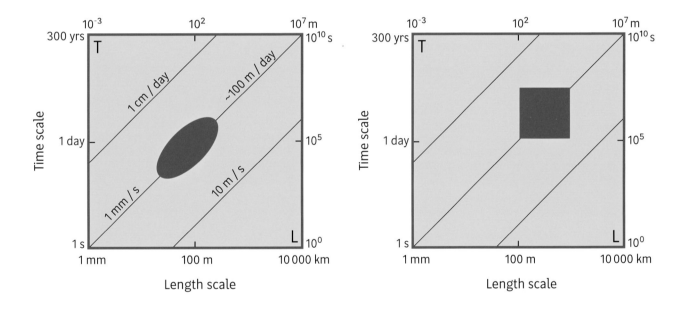

Stommel diagram: Length scales from 1 mm to 10 000 km. Time scales from 1 sec to 300 years. Ranges were selected to conform to what humans can experience directly. Left: Ellipse indicates the space-time scales of the underlying processes. The oblique lines reflect the characteristic velocities of the processes. Right: Square indicates the space-time scales (in terms of spacing and extent) of the sampling method. One of these diagrams is depicted on all photos in this book, depending on whether the photo represents a process or a sampling method. The ellipses and squares always have the same size for clarity.

Stommel's motivation for his diagram was the sampling design of oceanographic expeditions. As he pointed out, expeditions are scientific experiments that need to answer specific questions, so the space and time scales of the sampling need to be commensurate to the processes to be captured. If the sampling is too far apart (i. e. their spacing is too large) the small scale variability (in space and time) will fall through the cracks. If the sampling is too short (i. e. their extent is too small) the large scale variability will not be mapped. Finally, if the measurement volume (also termed the support or footprint) is too large, as is the case, e. g. with some satellite sensors, too much of the small scale variability will be smoothed out. In other words, we need to sample at the right space-time scale, otherwise we miss important information. These considerations will help in designing the strategy of explorations with a particular process in mind, similar to Stommel's strategy of deploying ships and buoys. The fantastic instrumentation depicted in this book all sample at different space and time scales and these are, again, visualized by small diagrams where the spacings and extents of the sampling are represented by squares. For example, a satellite sensor with pixel sizes of 100 m and 30 km coverage, and daily revisits over a period of one year would plot as a rectangle in the Stommel diagram as shown in the right panel on the previous page. Only processes with space-time scales within the rectangle will be captured by the sampling. In the example of the diagram, processes with characteristic velocities of 100 m/day will be captured, while faster (10 m/s) and slower (1 cm/day) processes will not be captured. Generally, the sampling scales plot as rectangles on the diagram but, for clarity, they have been plotted as same-sized squares in this book.

Over the years, the Stommel diagram has found its way to essentially all the geoscience disciplines and has been widely used both for visualising process dynamics across scales and for sampling design. The scale concepts have themselves voyaged across the geoscience disciplines.

The usefulness of the scale concepts extends beyond measurements to modelling and predictions. In fact there is a scale problem common to most geosciences: the samples are taken at small scales but we would like to make predictions at much larger scales, a problem termed upscaling. An example are the measurements of carbon fluxes by a flux tower that has a footprint (or sampling volume) of about one hundred metres across. Yet we are interested in the carbon flux of an entire continent. Another example is the measurement of the hydraulic characteristics of a soil that has a sampling volume of a few centimetres across. Yet we are interested in the water fluxes from entire river basins. Similarly, ice cores have a typical diameter of a decimetre, yet we are interested in the characteristics of an entire ice shelf. How do we get from these small sample scales to the scales at which predictions are required? How do we upscale?

The geoscience disciplines have developed a range of methods to do the upscaling. First, there is the statistical approach to relate, say, the soil properties across scales. Geostatistical methods based on spatial correlations are often assisted by auxiliary pattern data of surrogates of the variables of interest as can be obtained by remote sensing methods. The idea is to link the space scales through the observed patterns of the surrogates. The shape of the spatial correlation functions themselves is also of interest, in particular when they span many orders of magnitude. If the patterns are self-similar — or more generally if they are scale invariant — the correlations obey power laws, and (with some "squashing") the whole of the correlation function has the same shape as a part of it. There are more elaborate methods such as random cascades that exploit other aspects of scale invariance to create fractal patterns at different scales whose degrees of filling space can be quantified by hierarchies of fractional dimensions.

Alternatively, there are deterministic, spatially explicit models of the dynamics of the underlying processes. The incomplete data are fed into the models by methods termed data assimilation and the models themselves are used to fill the gaps between measurements. Often, a scale problem remains if the model cell size is bigger than the sampling volume. The methodological question then is how to get model parameters at larger scales that are consistent with the underlying model dynamics from the parameters at smaller scales. Effective parameters are one such way to obtain the consistency. Sub-grid parameterizations and closure relationships are other possibilities. There are fantastic modelling examples in this book that all address the scale issue in some way and simulate accurate patterns.

Even though the processes vary dramatically within the Earth system, there are lots of similarities of scale issues between the geoscience disciplines. In all cases we need to plan the sampling strategies and their space-time scales, be it through boreholes, in situ sampling or remote sensing methods. In all cases we relate the data to the underlying, incompletely sampled processes and we use the data to feed models to better understand the Earth system. As for every pattern we can see, there are many more patterns we cannot see and these we mean to capture with our models. There are therefore synergies between the geoscience disciplines we may want to exploit. We may want to learn from our sister disciplines about how they are testing their hypotheses, how they are conceptualising the underlying processes and how they deal with scale. The diverse set of examples in this book is intended to whet the appetite. Why not exploit the similarities between the geoscience disciplines? We may be amazed by the exciting results our colleagues have in store for us.

As the Earth and humankind continue the voyage through time we have arrived at a stage where the human imprint on the Earth system can no longer be overlooked. We have arrived at the Anthropocene. Most of our models start from the assumption that the geo-system can best be studied without human effects. There are of course very good reasons for leaving out humans, as they add enormously to the complexity. Yet, there is an increasing number and variety of patterns that can no longer be explained without integrating anthropogenic processes. The photos of this book bear testimony to it. A new thinking that revolves around the dynamic coupling of geo-processes and human action/reaction is therefore needed. Humans may no longer be treated as boundary conditions but should be seen as an integral part of the coupled human-nature system. As this coupling is taking centre stage more and more in Earth Science, the coupling between the geoscience disciplines also gets more important. Perhaps the concepts of scales can play a catalytic role in our endeavour of integrating our disciplines into Earth system science to better understand the interplay of processes across scales.

18
Top to bottom Zooming by factors of 2.9 into vertical sections of scale invariant simulated clouds.
Left Self-similar enlarged parts of structures are similar to the whole, so that the patterns do not change with scale.
Right A more realistic stratified but still scale invariant, cloud: enlarged parts of structures are similar to the whole if after each enlargement they are "squashed" by a factor of 1.6 in the horizontal; the patterns do change with scale. The top panels (left and right) are 5 km across, the bottom panels are 1 m across. In the Stommel diagram, the slope of the blue line is either 2/3 or 6/5 depending on whether the horizontal or vertical structures are used to define the size of structures (L).

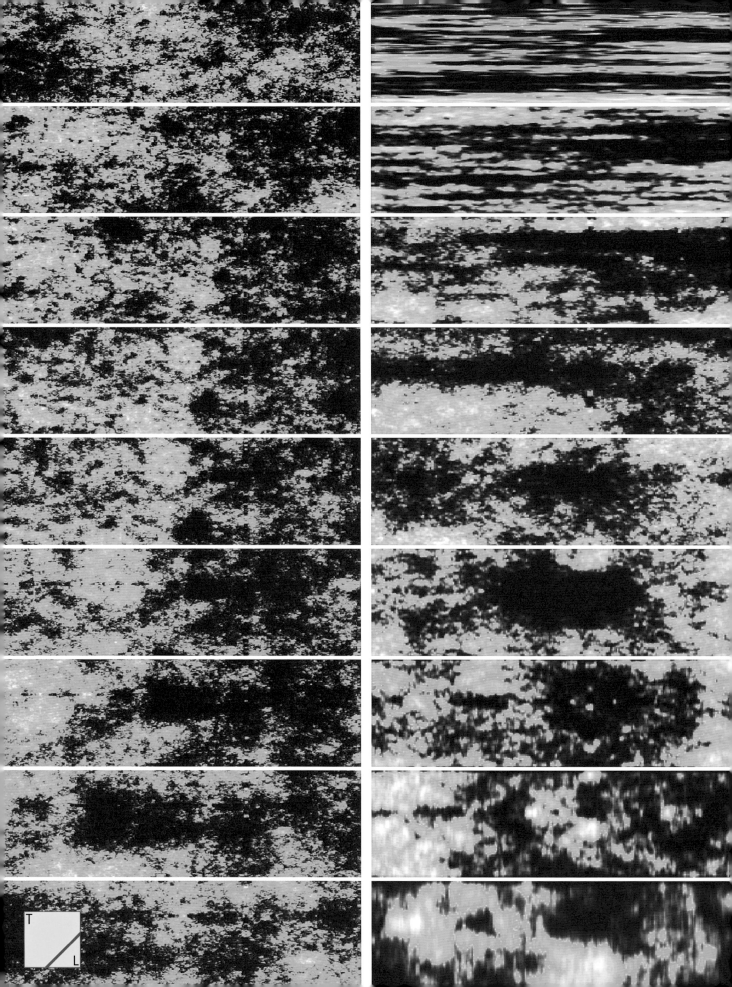

WHEN WILL I CEASE TO BE AMAZED AND BEGIN TO COMPREHEND? WHAT AM I? WHAT IS THE WORLD IN WHICH I LIVE?

GALILEO GALILEI

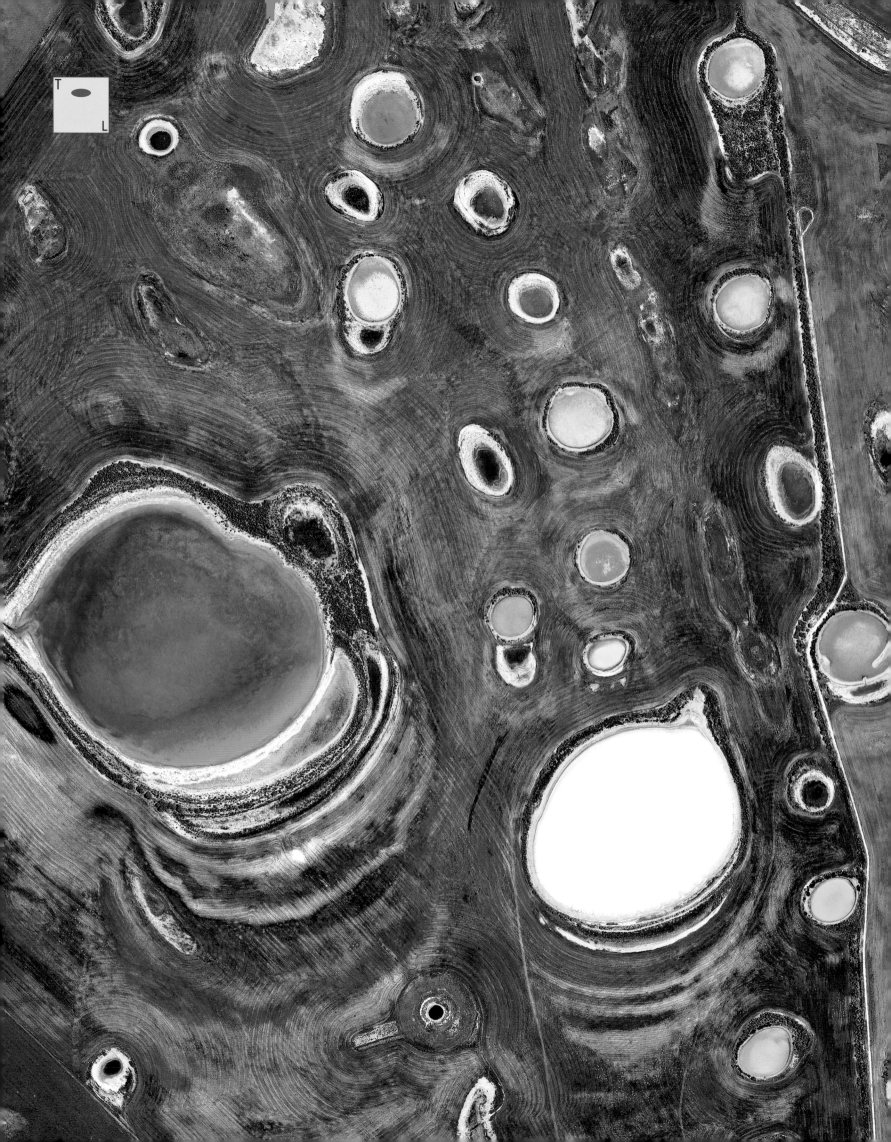

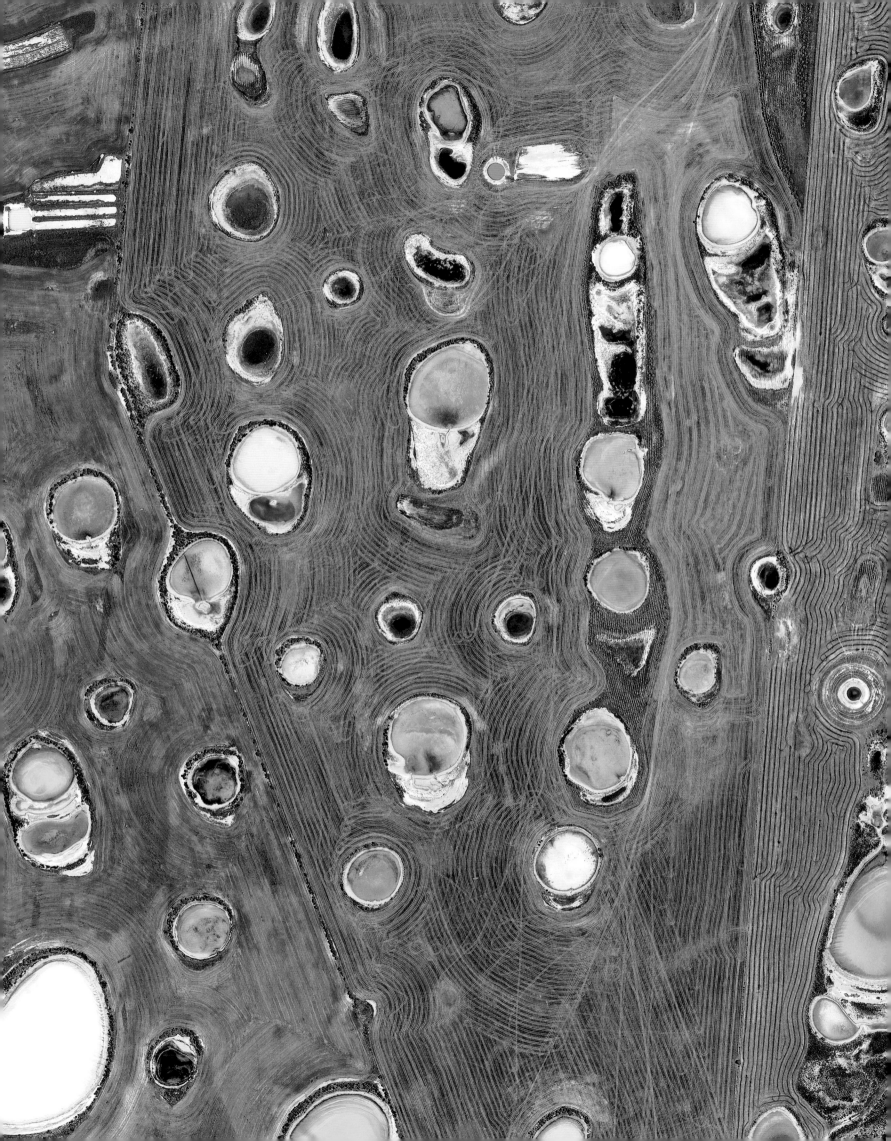

LIVING WITH A STAR

Ioannis A. Daglis, Christoph Jacobi, Mike Pinnock
Annales Geophysicae (ANGEO)

Life on Earth would be extremely difficult, if not impossible, if our planet were not positioned very precisely in the habitable zone in our solar system, colloquially known as the "Goldilocks zone". However, even positioned at this optimum distance from the Sun, the behaviour of our star has profound consequences for our lives. 'Living with a Star' is a challenge for a civilisation that has deployed technologies such as satellites and power grid systems that are vulnerable to particle energy emitted by our Sun. In recent decades, solar-terrestrial physics, the study of the interaction of the Sun with Earth, has been addressing the need to provide Space Weather forecasting to protect such technologies. This endeavour occupies a vast range of scales. The solar system is the largest complex system that mankind can study with in-situ observation. It involves dimensions ranging from the astronomical unit (1 AU = 150,000,000 km, the distance from the Earth to the Sun) to the radius of charged particle motions spiralling around magnetic fields which can be only a few centimetres. On the temporal scale, activity on the Sun varies on the "sunspot cycle" of 11 and 22 years whilst phenomena such as explosive energy events on the Sun and in near-Earth space require study on time scales of seconds.

Our Sun and near-Earth space region provide a natural laboratory for studying plasma physics processes. Plasma is the fourth state of matter and the most abundant matter state in the known Universe, showing properties unlike the other three states (gas, liquid and solid). The presence of free electrons and positive-charged ions, means that plasma can respond rapidly to variations in electric and magnetic fields.

The stunning false colour image (left) shows the complexity of plasma structures in the corona of the Sun. This image was taken by the NASA Solar Dynamics Observatory spacecraft and shows the temperature of the Sun's corona. The image reveals a huge amount of magnetic structure in the solar corona, but what is capturing scientists' interest are the occurrence of eruptions from the coronal surface, typified by the brown/pink structure in the centre of the image. Such coronal jets can hurl exceptionally dense balls of hot plasma towards the Earth.

From this activity on the Sun a stream of charged particles (the solar wind) is ejected into space with supersonic velocities, in addition to the electromagnetic radiation which supplies heat and light to the Earth. The planets sit as obstacles in this stream of particles. As a consequence, the Earth's magnetic field is distorted into a comet-like shape which is compressed on the dayside and dragged out into a long tail-like feature on the night side of the Earth. But our magnetic field is a leaky barrier to the solar wind, deflecting the majority of the particles away from the Earth but allowing some to enter. Typically, the solar wind power incident on the Earth's magnetic field is of the order of 10^{13} W. Depending on the precise conditions in the solar wind, power of $\sim 10^{10}$ W to 10^{12} W enters the Earth's magnetic cavity, the "magnetosphere".

The way in which that energy enters the magnetosphere and the consequences are the subject of much investigation. In the latest generation of satellite missions, arrays of spacecraft such as ESA's Cluster mission, are exploring plasma physics in our magnetosphere. Cluster is truly on a "Voyage of Scales", navigating the different regions of the magnetosphere, measuring the particles and the electric and magnetic field variations. Solar wind electrons enter the magnetosphere with energies of ~ 30eV (electron volts); electrons with energies up to 15 MeV are found in the trapped particle regions close to the Earth (the Van Allen radiation belts). Our magnetosphere is, in effect, a giant particle accelerator.

These energy transfer processes are invisible to mankind, though one of them gives us a spectacular illustration of what is happening — the aurora. The auroras are light emissions at 100—300 km altitude, triggered by energetic particles accelerated down to our atmosphere. The geometry of the Earth's magnetic field guides the particles into a circumpolar ring in both hemispheres. This phenomenon is seen on all planets having a magnetic field and an atmosphere — the illustration shows aurora on Jupiter. Whilst the circumpolar ring of light may be ~ 3000 km in diameter on the Earth, individual auroral forms reveal very fine scale detail that goes down to tens of metres. It is still a significant challenge to understand how the energetic particles, the Earth's atmosphere and the electric and magnetic fields interact to produce such fine scale features and then vary them, second by second, in location and intensity of light emission. Ground-based experiments, typified by the EISCAT radar systems (the European Incoherent

False colour image of the temperature of the Sun's corona where red is cooler, blue is hotter, from data collected by NASA's Solar Dynamics Observatory.

Scatter Scientific Association), can address such questions. Whilst satellites can make in-situ measurements of plasma parameters as they fly through the aurora, ground-based instruments can provide a more continuous measurement at a specific location. The two techniques are highly complementary. The operation of radio and optical instruments in both polar regions provides significant engineering challenges but such instruments are now being operated routinely at latitudes right up to the Poles.

Energy, however, can also propagate upwards from the lower atmosphere and modify the near-Earth space region. The Earth's atmosphere contains a rich variety of waves and tidal signatures which move energy around the planet. One illustration of this is shown in the NASA IMAGE satellite data. The bright, white spots in this picture show regions where the electron density in the Earth's upper atmosphere (the ionosphere) is being modified by upward propagating waves from intense thunderstorm activity over South America, Africa and Southeast Asia.

In summary, researchers have uncovered a complex series of energy transfer processes, originating from activity on the Sun, and then interacting with planetary magnetic fields and atmospheres. Such processes are of intrinsic interest, in particular to understanding the plasma physics processes that are ubiquitous throughout the Universe. However, an additional spur to this research is that we have technology systems that are either immersed in the energetic particle environment of space (e. g. satellites) or can be influenced by the energy released in phenomena such as the aurora (e. g. power grid systems with infrastructure near the auroral zone). Such systems can be damaged by "space weather" events. However, just as mankind has learned to use climate data and weather forecasting to adapt to and mitigate the effects of hazardous weather systems on our civilisation, we can do the same for space weather. Here lies the challenge.

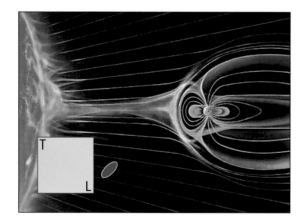

Schematic illustration showing how the solar wind from the Sun (left) distorts the Earth's magnetic field (right) into a comet-like shape.

ANGEO Editors-in-Chief

IOANNIS A. DAGLIS, *University of Athens (Greece); iadaglis@phys.uoa.gr*
CHRISTOPH JACOBI, *University of Leipzig (Germany); jacobi@uni-leipzig.de*
MIKE PINNOCK, *British Antarctic Survey, Cambridge (UK); mpi@bas.ac.uk*

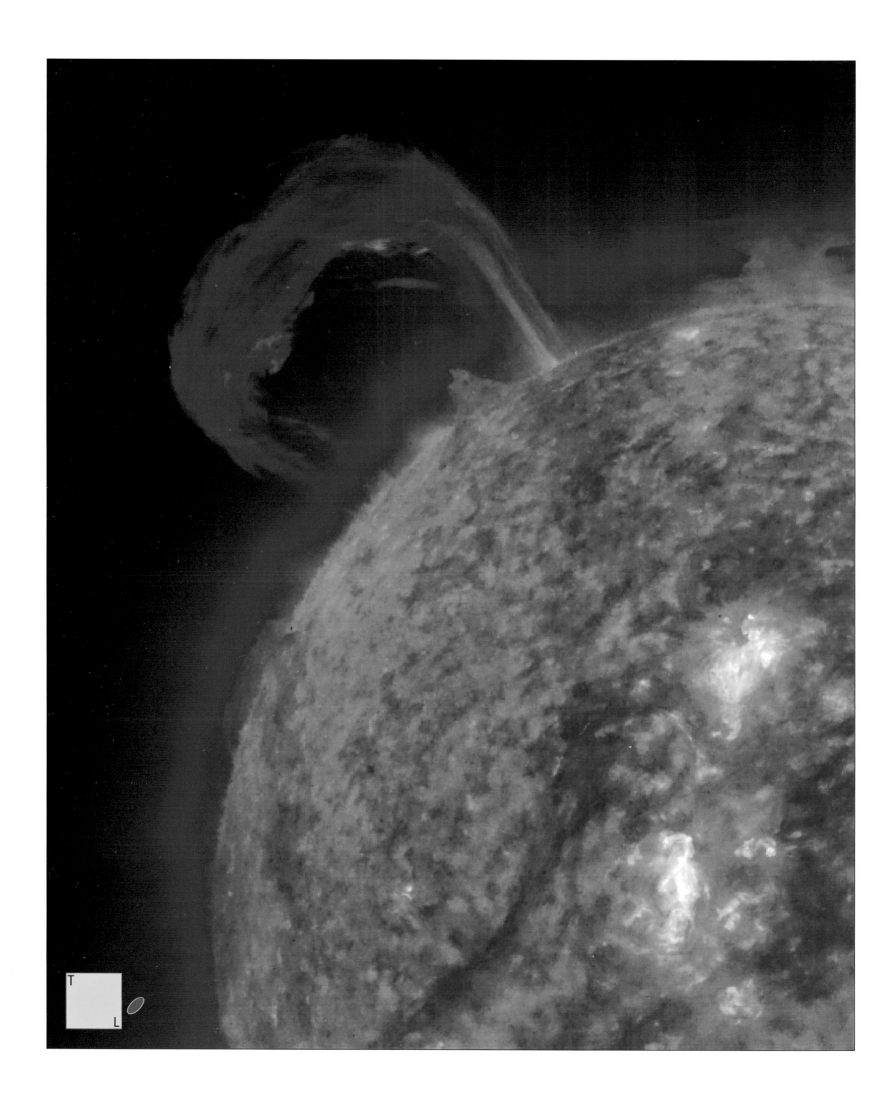

One of the EISCAT radar dishes photographed against a background of the northern lights. For scale size, note the person stood at the right-hand end of the radar structure.

Aurora on Jupiter. An artist's impression composed from spacecraft images. The circumpolar ring of light centred on Jupiter's north pole is approximately 35 000 km in diameter.

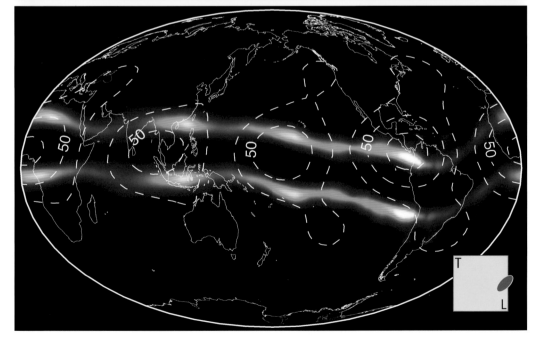

Image of ultraviolet light from two plasma bands in the ionosphere that encircle the Earth over the equator. The densest plasma areas (white) occur in the most active thunder storm zones. Image obtained by NASA's IMAGE satellite.

Left page:
An image of a portion of the Sun in extreme ultraviolet, showing a huge (many thousands of kilometres) solar prominence

In honor of SDO's fifth anniversary, NASA has released a video showcasing highlights from the last five years of sun watching
http://youtu.be/GSVv40M2aks
Aurora Borealis over Greenland and Iceland
http://vimeo.com/112360788

Laura Felgitsch, Hinrich Grothe
Atmospheric Chemistry and Physics (ACP)

Atmospheric processes extend over a vast range of temporal and spatial scales from less than nanoseconds and nanometres up to years and thousands of kilometres. Typical examples are chemical reactions and physical interactions of gases and aerosol particles at the small scale versus global circulation patterns of air masses in the troposphere and stratosphere at the large scale.

Clouds are prominent examples of atmospheric phenomena covering a wide range of scales. They extend over kilometres but consist of small water droplets or ice crystals in the micrometre size range, which form by condensation of water molecules on nanometre- or micrometre-sized aerosol particles serving as cloud condensation or ice nuclei. Depending on their microstructure and meteorological conditions, clouds may form different types of precipitation (rain, snow, hail) and play an important role in the hydrological cycle. Moreover, they influence the Earth's energy budget by scattering and absorbing sunlight and infrared radiation. Anthropogenic pollution can influence the atmospheric abundance and properties of aerosols and clouds, but the effects are quantitatively not well constrained and contribute substantially to the uncertainties of climate prediction. Thus, the interactions between aerosols and clouds are key elements of current atmospheric and climate research.

The pictures of this book chapter illustrate the multi-scale nature of atmospheric phenomena and research, focusing in particular on clouds and aerosols. One such phenomenon is Hurricane Linda, which was one of the strongest eastern Pasific hurricanes measured so far. Luckily Linda's eye never hit land and the storm remained mostly on open sea.

The convective clouds of individual thunderstorms (cumulonimbus) are much smaller but can still extend over dozens of kilometres horizontally and up to the tropopause near 15 km altitude. The formation and evolution of these and other types of clouds depend on a wide range of factors including meteorological conditions such as temperature and humidity as well as the abundance and properties of aerosol particles.

Cold clouds also contain snowflakes on a millimetre scale. The formation of such beautiful crystals is often triggered by aerosol particles serving as ice nuclei, which include biological particles like bacteria, fungal spores or pollen grains with diameters in the range of 1-100 μm. In the presence of ice nuclei, freezing can occur at temperature close to 0° C, while the freezing of tiny water droplets without ice nuclei requires temperatures as low as -40 °C. The temperature at which ice nuclei can induce the formation of ice depends on the chemical and physical properties of their surface, but the activity of ice nuclei is not yet well understood. Thus the formation of ice crystals, which is also important for the formation of precipitation, remains difficult to predict. These and other open questions of aerosol and cloud research are the reason why the IPCC (Intergovernmental Panel on Climate Change) ranks aerosol-cloud interactions as one of the most important uncertainties in the understanding and prediction of climate change.

A multitude of experiments have been developed to study and characterize ice nucleation and other atmospherically relevant processes under well controlled and reproducible conditions. Temperature- and humidity-controlled reaction chambers and flow tubes enable the investigation of aerosol and cloud particle ensembles on spatial and temporal scales of metres and minutes to hours. Other experimental approaches like levitation trapping, powder diffraction, and oil-matrix emulsions allow researchers to focus on the properties of particle ensembles or single particles on smaller spatial scales from centimetres to micrometres, but over a wider range of temporal scales from less than seconds to more than days. For example, high precision measurements of homogeneous ice nucleation rates in individual microdroplets levitated in an electrodynamic balance help to address the question of whether ice nucleation is initiated in the bulk liquid or near the surface of a droplet. Alternative methods carried out with droplets resting on different types of substrates require the consideration of droplet-substrate interactions that may interfere with the freezing process.

The methods applied in atmospheric chemistry and physics laboratories often resemble those in the fields of catalysis and materials science reaching down to the sub-nanometre scale. These include ultrahigh vacuum (UHV) investigations of molecular complexes, clusters, nanoparticles, and adsorbed films. Examples are the use of secondary ion mass spectrometry to investigate interactions between condensed H_2O, NH_3, and $HCOOH$ molecules,

Shining polar stratospheric cloud over Oslo. These clouds are located at 15—25 km altitudes. Due to the curvature of the Earth, they catch sunlight after sunset.

molecular beam studies of interactions between water molecules and water ice, and the surface-science experiments applied to study molecular-level surface properties. Several experiments that originally required UHV have also been adapted to higher experimental pressures, allowing for the investigation of water samples and cloud processes. Examples of such methods include electron microscopy, mass spectroscopy, reflection infrared spectroscopy, X-ray photoelectron spectroscopy, near-edge X-ray absorption fine structure, and molecular beam experiments.

New developments in laboratory experiments, field measurements, remote sensing and atmospheric modelling continue to open new pathways in the investigation and understanding of atmospheric processes on molecular to global scales, which are highly relevant for the understanding of atmospheric and climate change. Thanks to great efforts and achievements of numerous researchers, studies and methods, many scientific questions have been answered over the past decades and have helped to resolve issues like the ozone hole. However, many more issues relevant for air quality, climate and public health remain to be solved and this makes atmospheric science the exciting and important field it is today.

Top, middle **Pollen** grain in water ruptures and releases material; monitored by light and fluorescence microscopy.
Bottom **Freezing** water droplets (water in oil).

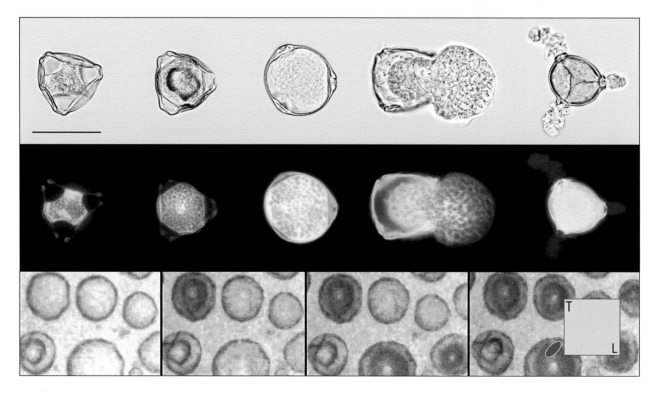

ACP Executive Editors

ULRICH PÖSCHL, *Max Planck Institute for Chemistry Mainz (Germany); u.poschl@mpic.de*
KEN CARSLAW, *University of Leeds (UK); k.s.carslaw@leeds.ac.uk*
THOMAS KOOP, *Bielefeld University (Germany); thomas.koop@uni-bielefeld.de*
ROLF SANDER, *Max Planck Institute for Chemistry Mainz (Germany); rolf.sander@mpic.de*
WILLIAM THOMAS STURGES, *University of East Anglia (UK); w.sturges@uea.ac.uk*

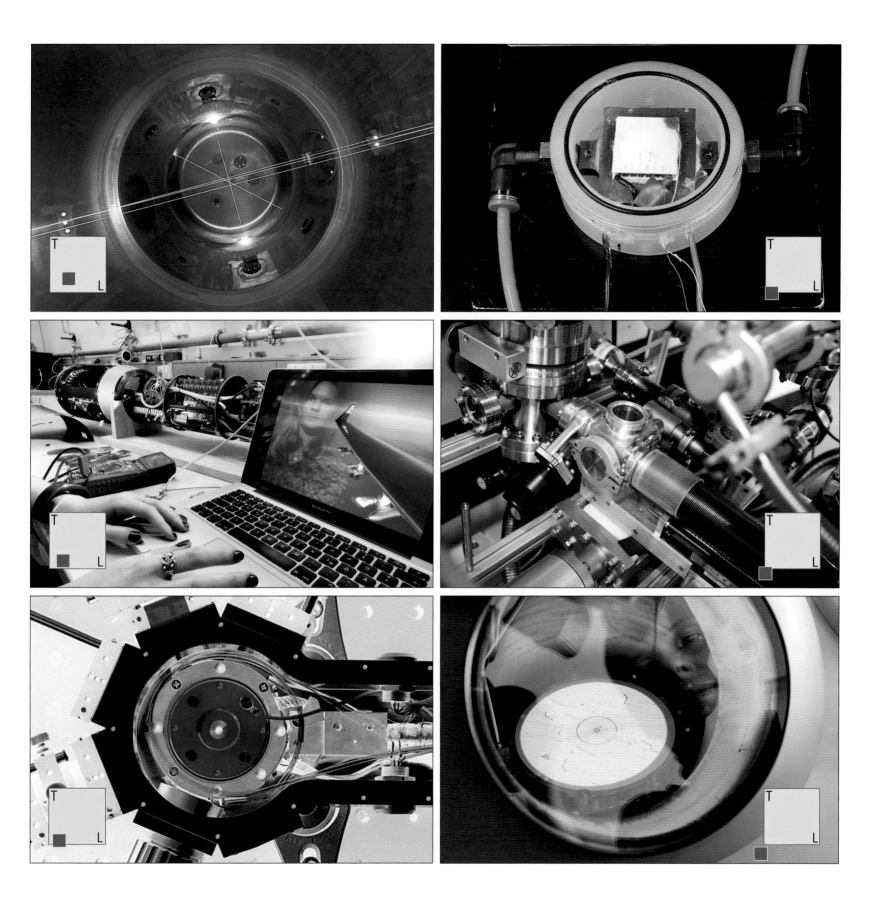

Left column down
Aerosol and cloud simulation chamber AIDA. Maintenance on the aircraft instruments PHIPS.
Electrodynamic levitator for examining individual droplets. All located at IMK – KIT (Karlsruhe).
Right column down
Peltier-cooling-stage for observation of ice nucleation of droplets in the μm range. High-vacuum
chamber and cryostat for taking vibrational spectra of molecules and clusters. A transmission
electron microscope displaying structures down to the picometre range.

A cumulonimbus cloud. If enough humidity is available they grow out of a conventional cumulus cloud until equilibrium is reached.

An ice crystal. Clouds consist of small water droplets or ice crystals (µm size range). Ice formation depends on temperature and humidity, as well as the presence of substances which trigger ice formation.

Pollen as an example of aerosols. Droplets or ice crystals need an active surface to allow water condensation.

Left page:
Hurricane Linda, an eastern Pacific hurricane, occurred in 1997. A Hurricane is a rotating storm with a low pressure centre accompanied by heavy rain and strong wind.

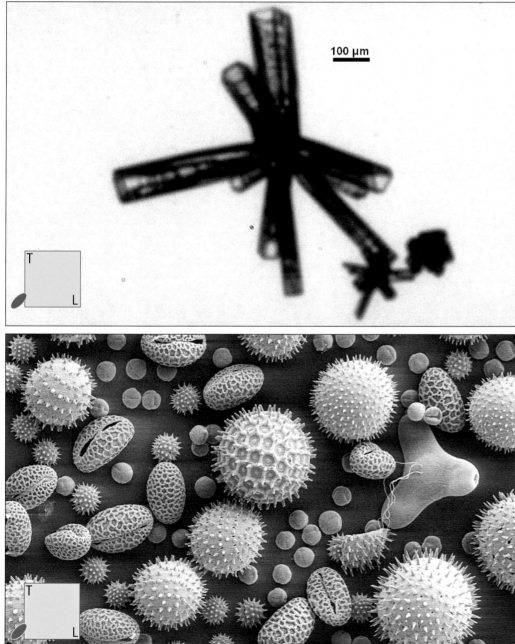

100 µm

Heterogeneous ice nucleation of supercooled water droplets
http://youtu.be/O0uwGlgkgfY
Aerosols: Airborne particles in Earth's atmosphere
http://youtu.be/uKTmTaukZHQ

SCALES AND ATMOSPHERIC MEASUREMENTS

Andreas Richter, Thomas Wagner
Atmospheric Measurement Techniques (AMT)

Atmospheric measurement techniques involve a broad range of methods from the development of in-situ and remote sensing techniques and instrumentation, laboratory and environmental chamber experiments, algorithm development to data intercomparison and validation and the quantification of measurement uncertainties.

The relevant scales of atmospheric processes and properties vary over many orders of magnitude. Even though early humans certainly had a clear grasp of the fact that weather and climate have a much larger force and scale than themselves, the full appreciation of the large scale patterns and connections of atmospheric phenomena was probably only possible with the first pictures and measurements from space. Over the last decades, these early photos of cloud systems have evolved to a global observational system of many parameters which feed the numerical models for weather forecasting as well as the scientific studies trying to understand the various processes of the atmospheric system. At the same time, more and more measurements are being made on medium and small scale processes to create a comprehensive picture of atmospheric processes at different scales and how they interact.

When designing a measurement of an atmospheric quantity, the first question to ask is always that of the scale of interest. Taking NO_2 concentration as an example, very different set-ups are needed when quantifying atmospheric levels at the exhaust pipe of a car, or in a street canyon of a city, or for a whole agglomeration, or on a country level. The difference in horizontal scale also results in differences in the natural vertical extension of the probing volume – while in a street, one would usually take point measurements of the NO_2 loading, a whole city is probably best viewed in the form of column amounts, integrated vertically over the boundary layer or the whole troposphere. The measurement volume also determines the range of values to expect – very large variability could be expected for reactive species at small scales, particularly close to emission sources, while much less variability is found at larger scales.

In the atmosphere, the space scales of the processes are intrinsically linked to their times through the transport velocities. When analysing measurements taken at a ground-based station, the spatial structures and scales of a phenomenon such as a high pressure system are mapped as temporal variations in the data depending on wind speed. The opposite is also true – the temporal evolution in the degassing of a volcano can cause visible spatial patterns of different sizes, for example, the distribution of aerosols or SO_2 in satellite images downwind of the crater.

Limited atmospheric life time also has an impact on the typical scales at which gases and particles are found in the atmosphere, which are relevant for observations. Considering a very short lived molecule such as the hydroxyl radical OH, transport can only provide mixing over very short scales and depending on local conditions, large gradients can be expected, for example, between illuminated air volumes and shaded areas. A trace gas with a long atmospheric lifetime such as CFC-11 on the other hand will be efficiently mixed in the atmosphere on continental and even hemispheric scales, leading to small gradients. As a consequence, the requirements for atmospheric measurements of the two species are very different – for a long-lived gas a few observational points globally might suffice while full characterisation of a short-lived constituent necessitates a large number of evenly spaced observations at considerably smaller scales and at much higher frequency.

The different scales of phenomena in the atmosphere are not independent of each other. On the contrary, they interact in many ways, transferring energy, constituents and properties between different scales. For example, large scale atmospheric motion such as high and low pressure systems dissolve over time, transferring energy and momentum through a cascade of smaller vortices and eddies until the kinetic energy is eventually dissipated to heat. At the same time, surface roughness from microscopic small scale structures to larger objects like trees, buildings or mountains determines surface friction which in turn has an impact on wind speed and direction on much larger scales. Another example is the formation of tiny aerosols in the atmosphere which can then act as condensation nuclei leading to the formation of clouds and impacting on both precipitation and the radiative balance not only locally but also on large scales. Similarly, polar stratospheric clouds (also known as mother of pearl clouds) are responsible for the efficient destruction of stratospheric ozone over Antarctica in October known as the ozone hole. Knowledge of the shape, composition and properties of individual sub-micron sized particles is needed and can only

Launch of a scientific balloon. Balloon borne observations provide precise measurements of the vertical distribution of ozone and other trace gases as well as meteorological parameters in the atmosphere.

be obtained by measurements on microscopic scales, while the effects on cloud formation and radiation can be observed on much larger scales, for example from satellites in space.

Another important example where microscopic phenomena control large scale processes is the exchange of trace species across phase barriers. Laboratory measurements in so-called wind channels in combination with observations of the microlayer structure of water surfaces are carried out to quantify trace gas fluxes through the air-water interface. For example, the efficiency of the air-sea exchange of CO_2 determines the rate at which the ocean can uptake and store the anthropogenically produced excess concentration of this greenhouse gas.

Remote sensing measurements, in particular if taken from satellites, create their own scale problem. In many cases, these observations are averaged over large volumes and areas. For example, satellite observations of tropospheric NO_2 have typical horizontal scales of hundreds of square kilometres and, in addition, integrate vertically over the troposphere. These scales are usually appropriate for mapping pollution on a regional level, and such measurements can provide global maps of NO_2 distributions which cannot be obtained by other observations. Remote sensing observations as an indirect method need validation by independent measurements. However, at the scales of the satellite observations, this is a challenge as no other measurement technique is available providing mean values over such a large volume. Averaging over many local measurements is needed, either using remote sensing from the ground or in-situ observations from air-borne sensors, sequentially probing the measurement column. As this is a time consuming process, and temporal scales of NO_2 variability in the atmosphere are relatively short, this can only lead to imperfect matching of measurement quantities. Clearly, the different scales of atmospheric phenomena can only be appropriately probed by different measurement approaches and instruments, creating a complex mosaic of measurements having diverse advantages and limitations but hopefully in the end providing a consistent picture of the atmospheric system as a whole.

Atmospheric measurements at different scales are often used for comparison with model simulations at similar scales. For example, satellite observations are well suited for the validation of global chemical transport models and general circulation models. Measurements from aircraft or satellite observations with high spatial resolution are compared to results of regional models. Regional models can be nested into global models with coarser resolution and can serve to bridge the gaps between observations at different scales (e.g. remote sensing and in-situ observations). Laboratory measurements of basic atmospheric processes (e.g. chemical and microphysical processes) are often compared to simple 1-dimensional box-models. While the scales on which phenomena in the atmosphere take place vary dramatically, it is the thorough and well-designed atmospheric measurement on all scales that provide the ultimate tool to test our understanding of chemical and physical processes in the atmosphere.

AMT Executive Editors

THOMAS WAGNER, *Max Planck Institute for Chemistry Mainz (Germany); thomas.wagner@mpic.de*
HARTWIG HARDER, *Max Planck Institute for Chemistry Mainz (Germany); hartwig.harder@mpic.de*
JOANNA JOINER, *NASA, Maryland (USA); joanna.joiner@nasa.gov*
PAOLO LAJ, *Grenoble University (France); laj@lgge.obs.ujf-grenoble.fr*
ANDREAS RICHTER, *University of Bremen (Germany); andreas.richter@iup.physik.uni-bremen.de*

Concentration pathlength (ppm-m)

| 0 | 500 | 1000 |

False colour image of SO₂ emissions from the Fuego volcano in Guatemala. The picture was created with a UV camera comparing measurements of scattered sunlight at two wavelengths having different SO₂ absorption strength. The SO₂ amount is shown, integrated along the light path with red colours indicating high values. Such imaging data can be used to estimate volcanic emissions of SO₂ on short time scales and provide a link between local in-situ measurements and large scale satellite observations.

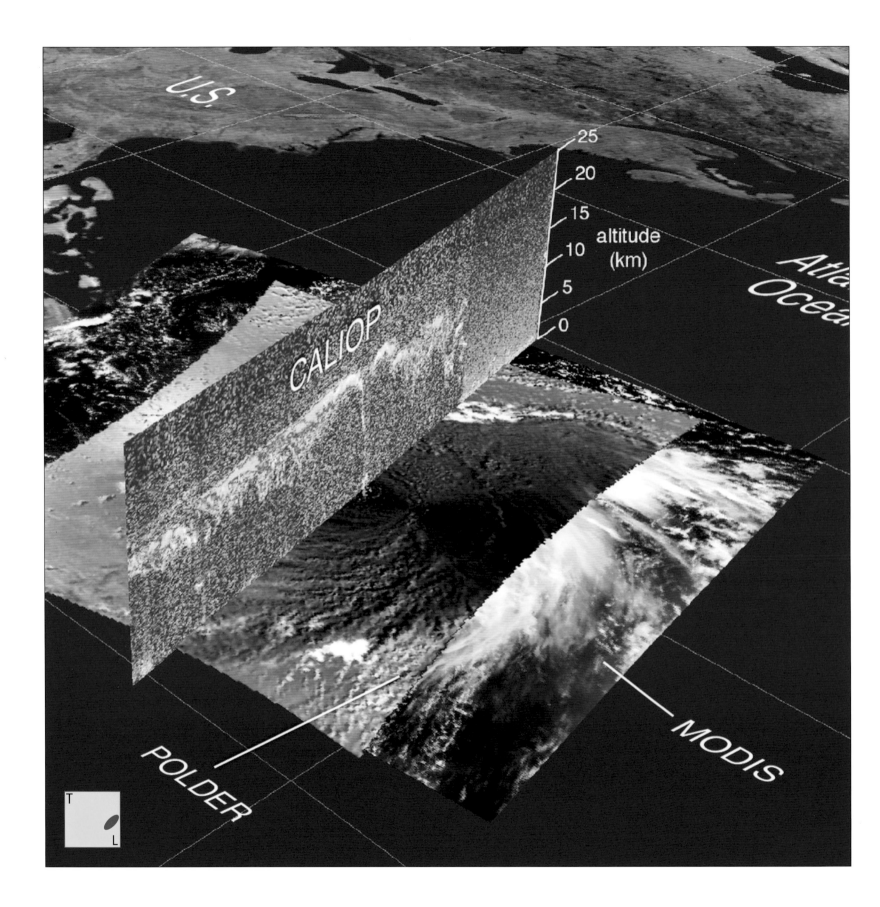

Illustration of measurements of three satellite instruments of Hurricane Bill in the Gulf of Mexico. The MODIS instrument provides high resolution images of cloud properties but only limited information on the cloud height. This is complemented by data from the CALIOP lidar which determines cloud heights at high vertical resolution but only for a narrow track through the storm. POLDER data show the polarized reflected sunlight which contains additional information on the cloud phase (liquid or ice).

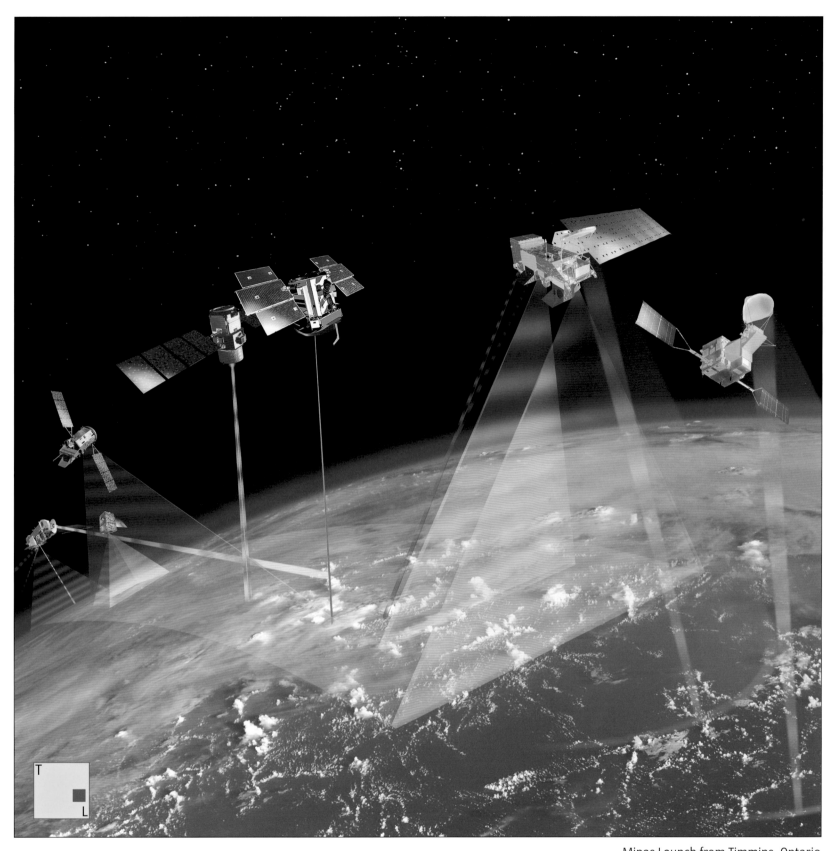

Mipas Launch from Timmins, Ontario
http://youtu.be/7w7anQOBEHw
Stratos: the Canadian Space Agency Stratospheric Balloon Program
http://youtu.be/Bq4CoQTqeQ0

Artist's concept of the A-Train, a series of satellites flying along the same orbit with time differences of a few minutes. While individual satellite instruments collect data on only a limited number of parameters, the combination of measurements from several platforms using different observation techniques provides a more complete picture of the atmosphere. Synchronizing the orbits, overpass times and viewing geometries of the satellites facilitates a synergistic use of their data products.

BIOGEOSCIENCES CONNECTING EARTH'S SPHERES FROM MICROSCOPIC TO GLOBAL SCALES

Michael Bahn, Katja Fennel, Jürgen Kesselmeier, Wajih Naqvi, Albrecht Neftel
Biogeosciences (BG)

Biogeosciences is focused on the interactions of marine and terrestrial life in and across the Earth's main spheres: the biosphere, atmosphere, hydrosphere and geosphere. These spheres are intricately connected and interact by exchanging energy and matter through biological, physical and chemical processes across a range of scales. Biochemically mediated reactions and transports across intracellular membranes occur at microscopic scales, yet they profoundly influence the composition of Earth's atmosphere and climate. Macroscopic properties, in turn, determine which biota and biochemical pathways are prevalent in various environments on Earth.

The wealth of interactions across spheres and scales leads to feedback mechanisms that can dampen or amplify natural and human influences. A striking example of a human perturbation to natural mass and energy flows on Earth is the steep rise in atmospheric carbon dioxide (CO_2) due to fossil fuel combustion, cement production and deforestation, which leads to global warming and acidification of the ocean. Because of human influences, we now refer to the period since the beginning of industrialization as the Anthropocene. The Anthroposphere, i.e. the sphere that is directly influenced by humans, permeates practically all other spheres on Earth.

Perhaps the prime example of microscopic processes fundamentally altering the global environment and the course of evolution is the oxygenation of Earth's atmosphere and ocean about two billion years ago. On the early Earth free oxygen was rare. The rise of photosynthesis in single-celled marine algae led to the accumulation of oxygen in atmosphere and ocean, and thus made the evolution of diverse multicellular life forms possible. In today's ocean the explosive growth of single celled algae, which occurs when light and nutrient conditions are favourable, can create accumulations of biomass that are visible from space. During senescence of these blooms, organic matter sinks to the deep ocean effectively sequestering carbon at ocean depths and mitigating global warming. As the surface ocean warms, the efficiency of carbon export to the deep ocean may slow, potentially creating a positive feedback that would amplify warming.

The ocean has taken up a large fraction of anthropogenic CO_2 from the atmosphere (estimated at a third to half of anthropogenic emissions) by a combination of sinking organic matter and physical equilibration processes, and continues to do so. The ocean has thus slowed man-made global warming, but at a price to ocean ecosystems. The increase in ocean inorganic carbon concentrations acidifies ocean water, making it more corrosive to calcium carbonate, which forms the solid structures of many marine organisms from select phyto- and zooplankton species (phytoplankton are small drifting algae and zooplankton their equally small predators) to bivalves (clams, mussels, oysters) and crustaceans (lobsters, crabs, shrimp). Perhaps the best-known early victims of ocean acidification and warming will be coral reefs — elaborate calcium carbonate structures built by symbiotic coral-algae associations that provide shelter to diverse tropical ecosystems.

On land, photosynthetic activity by vegetation results in the largest flux of carbon between the biosphere and the atmosphere, a large portion of which returns to the atmosphere as respired CO_2 within timescales of days to weeks. A smaller part of the assimilated carbon transiently remains in living biomass and ultimately feeds into the large carbon reservoir of soils, where the unseen majority of microorganisms transform soil organic matter, in turn sustaining the supply of nutrients for the vegetation.

Substantial alterations of biogeochemical cycles also include land cover changes and the introduction of non-native species impacting the structure and functioning of ecosystems. Industrial fertilizer production, a major advance that enables feeding the growing human population, has also massively perturbed the nitrogen cycle. Increased nitrogen deposition, which may take place over distances as large as several hundred or even thousand kilometres from the sources of production/emission, has consequences for the balance of macro- and micronutrients of organisms from the smallest microbe to the largest organisms on Earth. The laws of stoichiometry determine to which degree and how efficiently organisms compete for and process the elements, take up and store carbon and release greenhouse gases to the atmosphere. While increased nitrogen inputs to ecosystems can enhance the capacity of the vegetation to take up CO_2 from the atmosphere, they also speed up the microbially driven nitrogen turnover in the soil and stimulate the emissions of nitrous oxide, a potent greenhouse gas whose global warming potential is almost 300 times as high

An algal bloom on Lake Erie was observed from space in 2011 and is thought to result from runoff of agricultural fertilizers.

as for CO_2. On the continental scale, agricultural practices have recently been suggested to contribute to turning a small continental carbon sink into a net source of radiative forcing for both Europe and Africa. Along the nitrogen cascade, deposited nitrogen may also leach from terrestrial ecosystems and have downstream consequences for rivers, streams, lakes and coastal oceans.

Dairy products play a growing role in the nutrition of the expanding world population. Key to the growing importance of dairy products is the husbandry of ruminants. Ruminants are mammals that can acquire nutrients from grasses by bacterial fermentation processes in their first stomach — a complex bioreactor. The fermentation creates a range of gaseous byproducts including the potent greenhouse gas methane, which escape in large enough quantities to alter atmospheric greenhouse gas concentrations and thus climate.

Biosphere-atmosphere exchange measurements by the eddy covariance technique are a typical example of how to bridge the scale from individual point sources, the cows, to the whole grazing field, typical on the hectare scale. The application of eddy covariance over pastures is challenging due to the spatially and temporally uneven distribution of the grazing animals. Individual cow positions can be recorded by GPS trackers to attribute fluxes to animal emissions using a dispersion model that makes assumptions about the upwind sources of the fluxes. Meaningful interpretation of measured fluxes can only be gained if several disciplines work together, ranging from the development of instrumentation, to numerical modelling and animal physiology.

Another potential influence of animal husbandry is the release of amines into the atmosphere, which act as condensation nuclei that stimulate cloud formation thereby increasing Earth's albedo and slowing down global warming. The process interactions span scales from microscopic (bacterial fermentation and cloud condensation nuclei) to scales of metres (cows and their human handlers) to global (climate on Earth).

Critical questions for civilization are to what degree the biosphere can absorb greenhouse gases, tolerate warming and other environmental perturbations, and maintain livable conditions on Earth. During a period of rapid losses of biodiversity across the globe we need to understand how organisms and their interactions influence the fluxes of matter and energy across scales, and how these patterns and processes are altered by changing environmental conditions.

Biogeosciences research includes a wide range of perspectives from laboratory studies that quantify processes under controlled conditions, and field investigations that observe complex natural systems with feedbacks, to theoretical and numerical modelling studies that synthesize observations and allow a scaling up of local observations in space and time. The buildup of information resembles a puzzle with infinite pieces, but each new piece sharpens the contour of a steadily evolving picture of intricate patterns and interacting processes from microscopic to global scales.

BG Co-Editors-in-Chief

MICHAEL BAHN, *University of Innsbruck (Austria); michael.bahn@uibk.ac.at*
KATJA FENNEL, *Dalhousie University (Canada); katja.fennel@dal.ca*
JÜRGEN KESSELMEIER, *Max Planck Institute for Chemistry Mainz (Germany); j.kesselmeier@mpic.de*
S.W.A. NAQVI, *National Institute of Oceanography, Goa (India); wajihnaqvi@gmail.com*
ALBRECHT NEFTEL, *Agroscope ISS, Zürich (Switzerland); albrecht.neftel@agroscope.admin.ch*

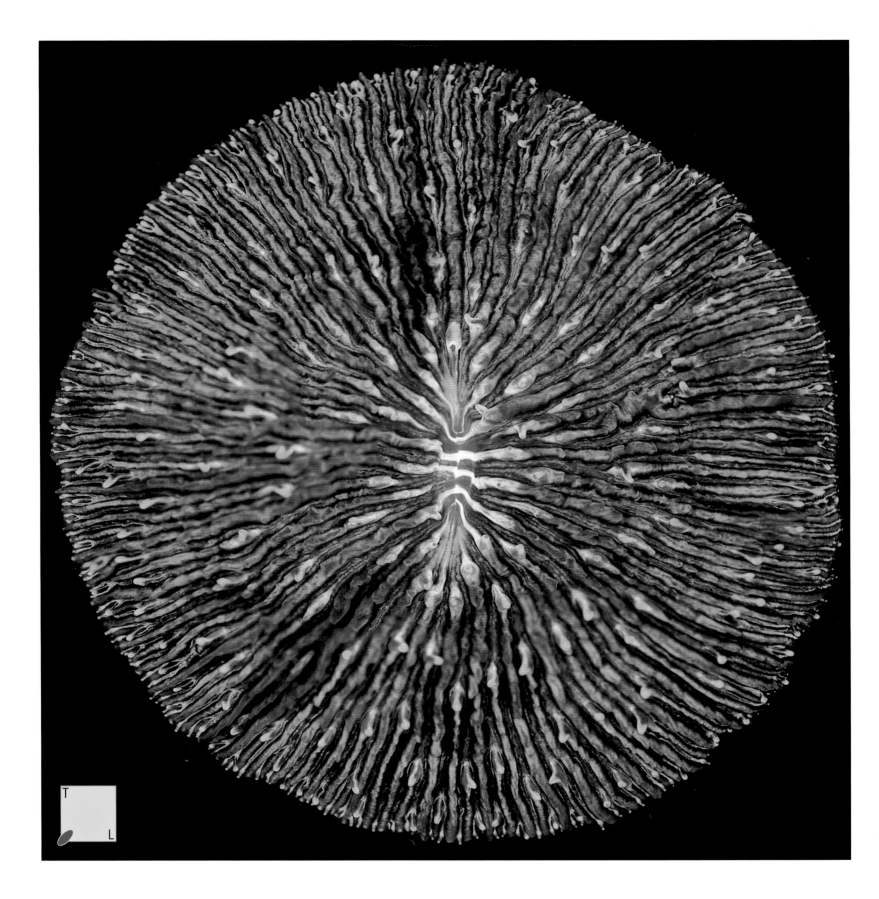

A large single coral polyp under filtered light. This form of chromatography is used to diagnose disease in corals. The mouth is the vertical slit in its center.

The Global Challenge of Ocean Acidification
http://youtu.be/5cqCvcX7buo

A garden of delicate coral in Kimbee Bay, Papua New Guiney, helps sustain local
fishermen. Kimbee reef is a hotspot of biodiversity.

SCULPTING THE SURFACE OF THE EARTH

Tom Coulthard, Jérôme Gaillardet, Frédéric Herman,
Niels Hovius, Douglas Jerolmack, Andreas Lang
Earth Surface Dynamics (ESurf)

Tectonic, river, glacial, wind, slope and coastal process have shaped the surface of our planet and made it into the environment we live in. Cities inhabit the flat lands of floodplains made by rivers, precipitous mountain ranges created by tectonic forces are sharpened by the erosive power of ice and glaciers, the rolling hills are gradually moulded by the erosion of soil and slopes, and at the coast land is made through the deposition of deltas and trimmed away by coastal erosion. The Earth's surface may seem static, but it is dynamic: and over a range to time and space scales the action of water, ice, gravity and the sea all combine to create our world. Ever wondered why mountains only grow so high? Or why floodplains are so gentle and fertile? All of this is a product of Earth Surface Dynamics.

Over millions of years the collision of tectonic plates leads to the growth of mountain ranges. Above a certain elevation the climate cools and the gradual but powerful action of ice and glaciers trims away the tops of these mountains and carves deep valleys at their base. The rock is then ground and transported by glaciers to lower elevations where rivers act as giant conveyor belts episodically transporting their load of sediment down slope. At even lower heights where gradients and potential energy drops, rivers deposit vast tracts of this material creating the large flat expanses of floodplains and at the coast deltas. The balance between these glacial, fluvial and slope processes leads to the variety of landscapes and landforms we see today. Where glaciers dominate, deep wide, U shaped valleys form. Where rivers lead in cutting into the landscape steep V shaped valleys evolve. Both of which are balanced by erosion and deposition on slopes — with the inexorable creep of material downslope, smoothing sharp gradients into to rolling hills. Climate affects all of these processes, providing water and snow to drive fluvial and glacial impacts, enabling soil to develop to further breakdown material once thrust up in mountain ranges. Where climate is drier the action of water reduces and sediment can be carried by wind, leading to the deflation of surfaces, the formation of sand dunes and sand seas, as well as the deposition of dust. The vast and fertile loess soils of China — that are eroded by and give the name to the Yellow river — were once part of the Himalayas and were transported East by wind. Furthermore, whilst climate can shape the landscape, the shape of the Earth's surface can also feed back and alter the climate — mountain ranges shift circulation patterns and uplift air prompting the release of rainfall and the dissolution of the eroded material will remove CO_2 from the atmosphere.

This complex mixture of erosive and depositional processes all operate over a wide range of time and space scales, ranging from the uplift of mountain ranges over millions of years to a burst of turbulence in a river causing a pebble to move downstream. This vast range of scales may at first glance seem to make deciphering how the Earth's surface evolves impossible. But, there is a great similarity in how processes operate over these different scales. For example, the powerful braided river Tagliomento shown in the first image operates in the same way as a small stream draining a beach. There is a statistical similarity between the shape of the channels and how they move and migrate. Similarly, the branch like network of a river network draining a continent has a physical and numerical correspondence to rills and gullies that may form on a slope or in a field after a rainstorm.

This similarity over scale allows researchers to use physical and numerical models to conduct experiments where we speed up time and reduce space to understand how — and at what rate our Earth's surface changes. For example, the image on page 54 shows a gullied and dissected landscape of the Badlands National Park, South Dakota, and the figure on page 52 shows how a numerical model can replicate the same shape and form. We can then see how fast this landscape may respond to wetter or drier periods in our Earth's history and how this translates into changes in the Earth's surface. Furthermore, these processes are not limited to Earth. The image on page 53 shows how large braided channels have formed on the surface of Mars at some point in its history. Titan, one of Saturn's moons has similar channels, but instead of lying in rock and sediment they are carved into frozen methane. Therefore, the dynamics we see and study on earth can be used to understand how other worlds are shaped. Understanding Earth Surface Dynamics is also vital for exploring our planet's resources. For example, over time river floodplains and deltas become buried under newer sediments and these layers build up to become aquifers — stores of water, oil and gas. Knowing where and how these form have enabled humans to extract the resources we use to build our world. Yet, our actions also have an impact — giving rise to the geological age of the Anthropocene. How we change the climate will have a direct impact on how glaciers, rivers, wind and coastal processes continue to shape the world in the future.

The braided channels of the Tagliamento river in north-east Italy,
that flows from the Alps to the Adriatic Sea.

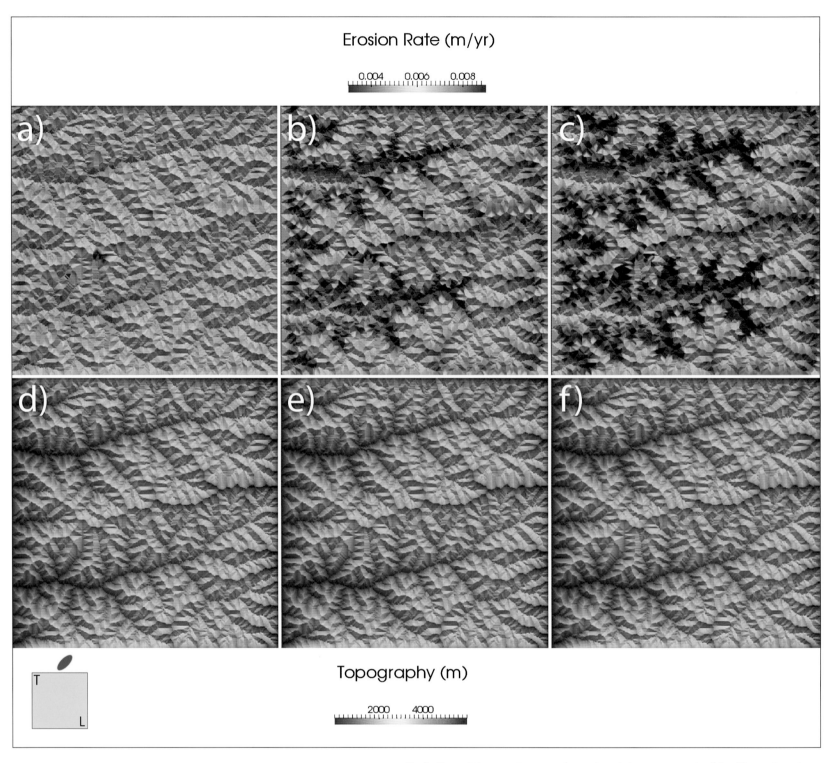

Erosion Rate (m/yr)

0.004   0.006   0.008

a)   b)   c)

d)   e)   f)

T

L

Topography (m)

2000   4000

Evolution of the erosion rate (a − c) and the topography (d − f) predicted
by the FastScape landscape evolution model showing the propagation of
waves of erosion through a landscape during a precipitation cycle.

ESurf Editors

TOM COULTHARD, *University of Hull (UK); t.coulthard@hull.ac.uk*
JÉRÔME GAILLARDET, *University Paris Sorbonne Cité and CNRS (France); gaillard@ipgp.fr*
FRÉDÉRIC HERMAN, *Université de Lausanne (Switzerland); frederic.herman@unil.ch*
NIELS HOVIUS, *GFZ GeoForschungsZentrum Potsdam (Germany); hovius@gfz-potsdam.de*
DOUGLAS JEROLMACK, *University of Pennsylvania (USA); sediment@sas.upenn.edu*
ANDREAS LANG, *University of Liverpool (UK); lang@liverpool.ac.uk*

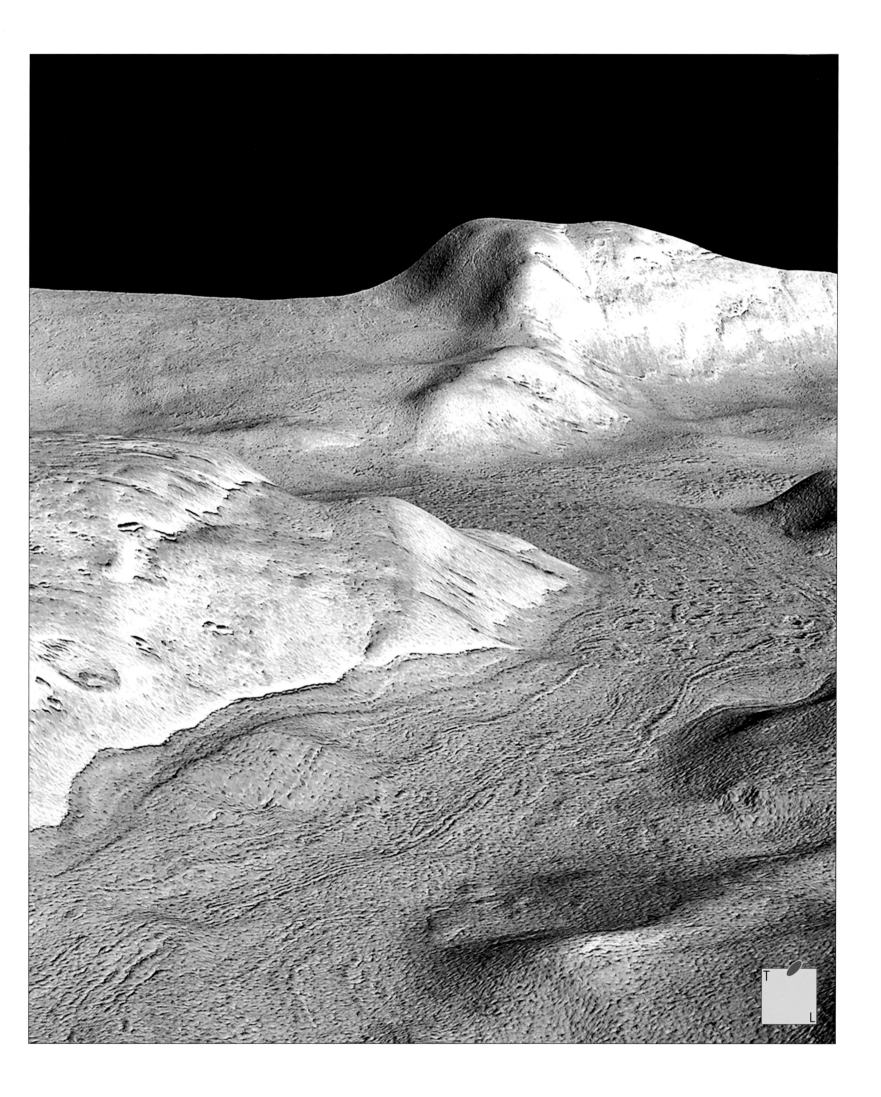

Erosional response of an actively uplifting mountain belt to cyclic rainfall variations (See Page 52)
https://youtu.be/gKifVnFwjEY

The striking landscape of the Badlands National Park, South Dakota.

EARTH SYSTEM DYNAMICS:
"THE WHOLE IS MORE THAN THE SUM OF ITS PARTS" (ARISTOTLE)

Somnath Baidya Roy, Axel Kleidon, Valerio Lucarini, Ning Zeng
Earth System Dynamics (ESD)

The planet Earth is a complex system consisting of multiple spheres such as the atmosphere, biosphere, cryosphere, hydrosphere, pedosphere, lithosphere, and the solid Earth. These spheres are tightly linked, resulting in a multitude of interactions and feedbacks that span the scales ranging from individual molecules to the planetary system. Because of these tight linkages, the behavior of the whole Earth system is more than the mere sum of the behavior of the component spheres. In addition to natural phenomena, the behavior of the Earth system is also influenced by human activity such as increases in atmospheric greenhouse gases and aerosols, and land use/land cover change due to deforestation, agriculture, urbanization, and utility-scale energy farms. In fact, the intensity and extent of human activity has reached such a level that the part of the environment influenced by humans is now considered a separate sphere called the anthroposphere that interacts with the other spheres. A systems approach is therefore essential to understand the Earth system and its response to global change, be it natural or anthropogenic.

While the processes that occur within each of the spheres appear to be unique, they all follow the fundamental laws of physics, so that we can expect a basic similarity between the processes in the different spheres. In addition to the internal dynamics that occur within the spheres, they interact with each other, continuously exchanging mass and energy amongst themselves. The uptake of carbon dioxide by the biosphere, for instance, is associated with the removal of carbon dioxide from the atmosphere, coupling the atmospheric composition to the dynamics of the biosphere. Due to this coupled nature of the Earth system, the processes that are purely internal to the spheres are strongly influenced by the exchanges at the boundaries, and thus by process that are external to the spheres. In this way, the overall behavior of the Earth system is shaped by the individual components as well as the interactions among the spheres.

Earth system processes and their interactions occur at a wide range of space and time scales. Atmospheric transport, for instance, spans molecular scale diffusion to planetary waves in the atmosphere. The Earth's climate provides an outstanding example of a high-dimensional forced and dissipative complex system, whose evolution is governed by Navier-Stokes equations in a rotating frame of reference complemented by the complex thermodynamical proper-ties of the multiphase geophysical fluids. The dynamics of such a system is chaotic, so that there is only a limited time horizon for skillful prediction. The challenge in predicting the system arises out of the vast range of spatial and temporal scales, the difference in the properties of the various components of the Earth system and the coupling mechanisms. For example, incoming solar shortwave radiation at the surface varies at short timescales of the order of minutes due to changes in cloud cover and aerosols to long timescales of hundreds of thousands of years due to changes in orbital eccentricity. On even longer time scales that span the Earth's 4.5 billion years of history, changes in solar luminosity, plate tectonics and greenhouse gases interact to create a habitable environment suitable for life. Due to the tight coupling in the Earth system, extremely small-scale, short-lived processes and very large-scale, long-term processes influence each other, even when they occur within different spheres of the system.

The connection between different scales are not readily apparent if we look at one phenomenon in isolation. It be-comes clearer when we take a holistic look at the entire system and take an interdisciplinary approach. Nowadays, the use of coupled Earth system models that simulate and predict the behavior of the whole system allows us to take such a holistic look at the whole Earth system.

One example for a process that spans a large range of scales is the exchange of energy, moisture and gases between the terrestrial biosphere and the atmosphere. The exchanges occur through stomata that are microscopic pores on a surface of leaves. The rates of exchange depend on a number of plant physiological, atmospheric, hydrologic and pedospheric characteristics that operate at different scales. The magnitude of the fluxes depend on the type and surface area of the leaves. For instance, transpiration from a leaf in a coniferous tree is much lower than that from a broadleaf tree. The structure of the tree canopy that governs the distribution of leaves on a tree is also a factor. Because stomata open to facilitate photosynthesis, fluxes are typically lower for leaves on lower branches that get less sunlight. Moreover, the small-scale variability of heat, moisture and carbon dioxide above, within and below the canopy due to the complex air flow inside a canopy also play a critical role. The fluxes also depend on soil mois-ture availability in the root zone that is governed by small scale variability in soil type and texture, as well as elevation and slope. At the larger end of the spectrum, the fluxes are governed by weather and climate phenomena that determine the spatio-temporal pattern of heat, moisture and gases in the atmosphere. Thus, the heat, moisture and gases that are exchanged through the stomata at the microscopic scale shapes the exchange of these substances between the land surface and the atmosphere, thereby playing a vital role at the exchanges at the regional to continental scale.

Another example for the need of a systems approach that spans a range of scales are the impacts of human-driven land use change through deforestation and agricultural expansion on weather and climate. Land-atmosphere interactions in the aftermath of deforestation involve phenomena at a wide range of scales, including stomatal exchange, atmospheric convection and thunderstorms (~10 km), catchment run-off and river flow (10—100 km) and monsoonal circulations (~1000 km). At smaller scales, sharp gradients between the forests and the clearings can create mesoscale circulations that affect the spatial dynamics of heat and moisture in the atmosphere. Even though sharp land cover gradients may occur naturally, the frequency is higher in anthropogenic land use patterns such as the fishbone deforestation in Brazil or the radial deforestation pattern in Bolivia. Similar effects are also observed with irrigated agriculture that often demonstrate an organized spatial structure with sharp boundaries. The effect is most pronounced in arid regions where the soil moisture differential between the irrigated farmlands and the surroundings are very pronounced. However, the effect of deforestation is not only felt locally but also in distant regions. Very large scale deforestation such as that in Amazonia can affect regional and possibly global weather and climate patterns. In the long term, such changes in weather and climate go on to exert feedbacks on the land cover in many different ways. They accelerate or dampen the vegetation regrowth rate and may even determine the species composition of the recovered forest. Because land use practices change over time, the nature and magnitude of the effects also evolve in time, thus linking near-instantaneous stomatal response to long-term climate response that may take decades to unfold.

Scale interactions in the Earth system is also evident in climate change that is mainly caused by the increase of greenhouse gases from anthropogenic fossil fuel emissions and land use change. Underlying the long-term increase in atmospheric background concentration of carbon dioxide (CO_2) is the intricate pattern of how the anthropogenic emissions from individual sources are collectively swept into swirls of weather patterns around the world. Most of the fossil fuel emissions come from population centers such as the world's major metropolitan areas in North America, Asia and Western Europe. These emissions are transported and mixed first in the Northern Hemisphere, then slowly diffuse into the Southern Hemisphere. In contrast with this nearly steady carbon emission source, the plants, especially during the summer growing season of the Northern Hemisphere, take up even more CO_2 from the atmosphere than the amount of fossil fuel emissions. This CO_2, however, is respired back into the atmosphere slowly during the dormant season, leaving only a small fraction that is sequestered on longer timescales as a net carbon sink for the fossil fuel emissions. The locations and nature of this so-called 'missing' carbon sink is still a major scientific mystery that involves scales ranging from individual plants to the planetary system.

The examples above clearly demonstrate the complex, nonlinear, multiscalar nature of the Earth system that, by virtue of the interactions between its component spheres, is much more than a mere sum of its parts. A systems approach that integrates knowledge from many different disciplines is required to understand and predict the behavior of the Earth system and its response to human activity.

ESD Chief Editors

AXEL KLEIDON, *Max Planck Institute for Biogeochemistry Jena (Germany); akleidon@bgc-jena.mpg.de*
VALERIO LUCARINI, *University of Hamburg (Germany); valerio.lucarini@zmaw.de*
NING ZENG, *University of Maryland (USA); zeng@atmos.umd.edu*

Biosphere-atmosphere exchange of moisture occurring through microscopic pores on leaves called stomata (top) depends on many factors. Leaves of broadleaf species (middle) with higher surface area transpire more than coniferous leaves. In tropical forest canopies (bottom), the leaves at the top tend to release more water into the atmosphere than leaves at the bottom where photosynthesis rates are lower due to lack of adequate sunlight.

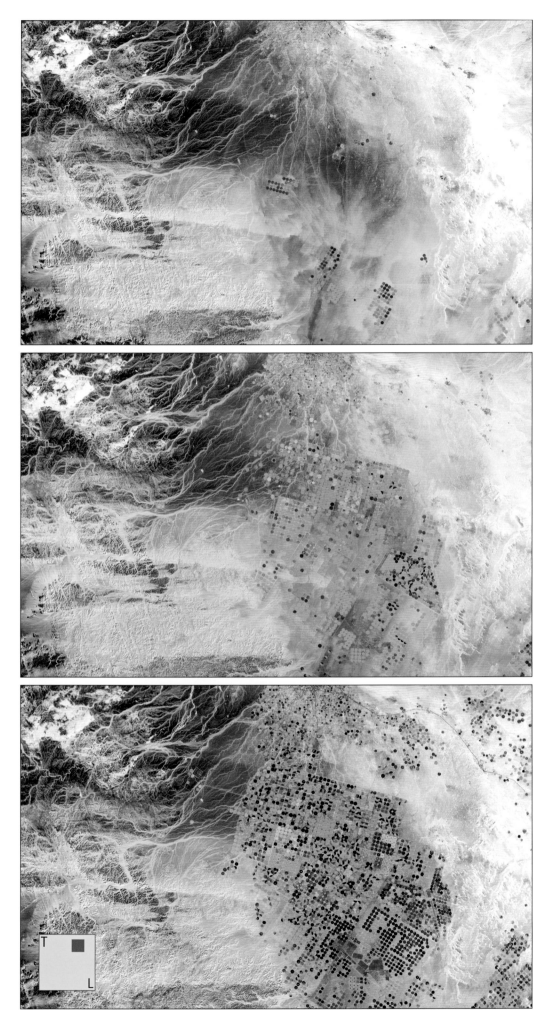

Expanding croplands
in the Wadi As-Sirhan Basin, Saudi Arabia, seen in a series of false-colour images for 1991 (top), 2000 (middle) and 2012 (bottom) from LANDSAT. The agricultural fields are about one kilometre in diameter and use center-pivot irrigation.

Page 110
Carbon dioxide patterns in the atmosphere on 27 April 2006, during the spring maximum in the Northern Hemisphere from the Nature Run by the Goddard Earth Observing System Model, Version 5 (GEOS-5).

SCALES IN ATMOSPHERIC REMOTE SENSING INSTRUMENTS, MEASUREMENTS AND SCIENTIFIC PRODUCTS

Jothiram Vivekanandan, Ari-Matti Harri, Håkan Svedhem
Geoscientific Instrumentation, Methods and Data Systems (GI)

In assessing the future trajectory of Earth's climate, scientists rely on climate models to predict potential changes to our planet. One of the largest remaining scientific uncertainties critical to improving climate model capabilities is the proper understanding and treatment of cloud processes. The Inter-government Panel on Climate Change (IPCC) has repeatedly stressed the adverse impacts on weather, sea level, food production and different habitats in response to small increases in Earth's temperature. These impacts can be predicted with greater confidence if the representation of clouds in climate models is improved. This is not a simple task since atmospheric clouds differ in intensity, shape, size and duration, ranging from small cumulus to tropical cyclones. Climate models have to accurately reflect spatial scales on the order of metres to thousands of kilometres, and temporal scales of minutes to days for representing various forms of clouds.

Clouds play various roles in the Earth's atmosphere. For example, puffy cumulus clouds reflect incoming solar radiation, which has an overall cooling effect whereas high altitude, wispy cirrus clouds trap outgoing radiation from the Earth's surface, warming our atmosphere. Tropical cyclones consist of a mix of clouds and thunderstorms and originate over tropical or subtropical oceans. They produce heavy rainfall, lower the sea-surface temperature in their vicinity, and maintain the global heat balance by transporting warm tropical moist air to the mid-latitudes and polar regions.

Aerosols play a major role in the formation of clouds. The interaction between aerosols and clouds is currently treated very crudely in climate models. A number of observational studies showed that aerosols affect cloud formation and evolution, as well as the timing and magnitude of precipitation. Models and field observations have shown that an increase in human-made aerosols can ultimately lead to higher concentrations of cloud drops, which in turn enhance a cloud's reflective properties, thereby leading to a cooling effect and affecting precipitation. Precipitation is the primary mechanism for transporting water from the atmosphere back to the land and ocean surface. It is interesting to note that a drop of water spends an average of eight days in the atmosphere before falling back to the Earth.

In order to reduce the uncertainty in climate model forecasts, detailed measurements are necessary for accurate representation of cloud microphysics in climate models. Cloud microphysics fields include types of atmospheric particles such as aerosol, ice, liquid, mean size, and concentration. Retrieved microphysical estimates from remote sensing instruments, namely, lidars and radars, are used for estimating mean characteristics of cloud microphysics. They are also used to evaluate a climate model's predictive skill, i.e., how well a climate model simulates 'present climate', which is an important part of evaluating a climate model's prediction of global warming. It should be noted that evaluating a climate model's performance using measurements is a daunting task since temporal and spatial scales of aerosols and clouds span many orders of magnitude, and the measurements should be compatible for inferring corresponding temporal and spatial scales of aerosols and clouds.

Remote atmospheric measurements from lidars and radars are used to estimate a wide range of spatial and temporal scales of aerosol layers and clouds. Radars detect micron size cloud liquid, cloud ice particles, raindrops, and large cm size ice particles. Lidars are often used to determine thin-cloud regions and aerosol layers of the atmosphere as they detect reflections of objects such as air molecules, aerosols, ice crystals, and water droplets in clouds. Lidars emit shorter wavelengths than radars, so they can detect atmospheric particles in the micron to millimetre size range. Lidar and radar together give a more complete picture of the atmospheric particles of sizes that range from micron to centimetre sizes and these particles are the primary constituents of aerosol layers and clouds.

Lidars and radars are active atmospheric remote sensing instruments. The sensitivity of these instruments for detecting atmospheric particles is proportional to the concentration and mean size of the particles. Since the spatial and temporal structure of aerosols, clouds and precipitation span a wide range of scales, these instruments are built to measure backscatter signals across those scales. Atmospheric particle sizes and concentrations detected by these instruments span many orders of magnitude.

This eye safe atmospheric lidar was developed by the Electro-Optical Systems Laboratory (EOSL) of the Georgia Tech Research Institute in Atlanta as a training and research instrument for undergraduate women at nearby Agnes Scott College.

This article provides a brief description of how remote sensing instruments are capable of detecting atmospheric particles, whose concentration and size vary over many orders of magnitudes. Examples of scientific products derived from lidar and radar measurements that have the potential for evaluating climate models are also presented.

Atmospheric remote sensing instruments transmit electromagnetic energy at various frequency bands between visible and microwave wavelengths. Visible frequency corresponds to 0.5 μm wavelength, and the lower end of the microwave frequency corresponds to 10 cm wavelength. Thus the range of transmit frequencies and wavelengths of remote sensing instruments span four orders of magnitudes. Energy transmitted at these wavelengths travels at the speed of light (3×10^8 m s^{-1}). In one millisecond the transmit and receive signals covers a range of 150 km. Minimum and maximum detection ranges of these instruments vary between 10 m and 300 km. Since the spatial structure of aerosol, clouds, and precipitation range between metres and kilometres, these instruments are designed to detect a wide range of temporal and spatial scales.

The transmitting energy pulse length determines the spatial resolution of a measurement. Shorter pulse lengths correspond to finer spatial resolution. Typically the transmitting pulse lengths vary between 25 ns and 1000 ns for resolving 1 m to 100 m scales. In order to cover a large volume and obtain a fine spatial resolution of atmospheric features, these instruments are designed to transmit the maximum allowable power. Radars are capable of pulsing mega watts of power at microwave frequencies, whereas lidars can transmit only kilowatts of power at visible wavelengths due to inherent engineering limitations and constraints for transmitting larger amounts of power. Even though transmitting power from lidars and radars differ by three orders of magnitude, they are sensitive enough to detect aerosol, clouds, and precipitation as described below.

When the transmitted electromagnetic wave from lidars and radars encounters discontinuity in space due to obstacles such as aerosols, clouds and precipitation, the energy is reflected back to the instrument. The amount of energy reflected back to the instrument is proportional to the number concentration and average backscattering cross section of the particles in the scattering volume. Concentration of micron size aerosols ranges between 10^8 and 10^{11} per cubic metre (m^3), whereas cloud and precipitation particles range between 1000 and 0.1 per m^3. The scattering cross section of aerosols is five orders of magnitude smaller than precipitation size particles, yet lidars detect aerosols because the typical concentration of aerosols is many orders of magnitude larger than concentration of cloud and precipitation particles. In the Rayleigh scattering regime, scattering cross section of a particle is proportional to fourth power of frequency. Therefore, as the transmit frequency increases by a factor of 10, the radar cross section of a particle increases four orders of magnitude. This large increase in the scattering cross section as the function of frequency enables detection of micron sized particles even though transmit power at visible frequencies is three orders of magnitude smaller than at microwave frequencies. Typical examples of measurements of lidars and radars and also retrieved atmospheric particles are shown on the next pages.

In summary, weather and climate spatial scales span seven orders of magnitude and their temporal scales range between seconds to years. Lidars and radars are best suited for detecting aerosols, cloud droplets, cloud ice and also large ice particles and raindrops. Detection of these atmospheric particles lead to objective depiction of clouds at a range of temporal and spatial scales that are necessary for improving predictions of climate change.

GI Executive Editors

JOTHIRAM VIVEKANANDAN, *National Center for Atmospheric Research, Colorado (USA); vivek@ucar.edu*
ARI-MATTI HARRI, *Finnish Meteorological Institute, Helsinki (Finland); ari-matti.harri@fmi.fi*
HÅKAN SVEDHEM, *ESA, Noordwijk (Netherlands); hsvedhem@rssd.esa.int*

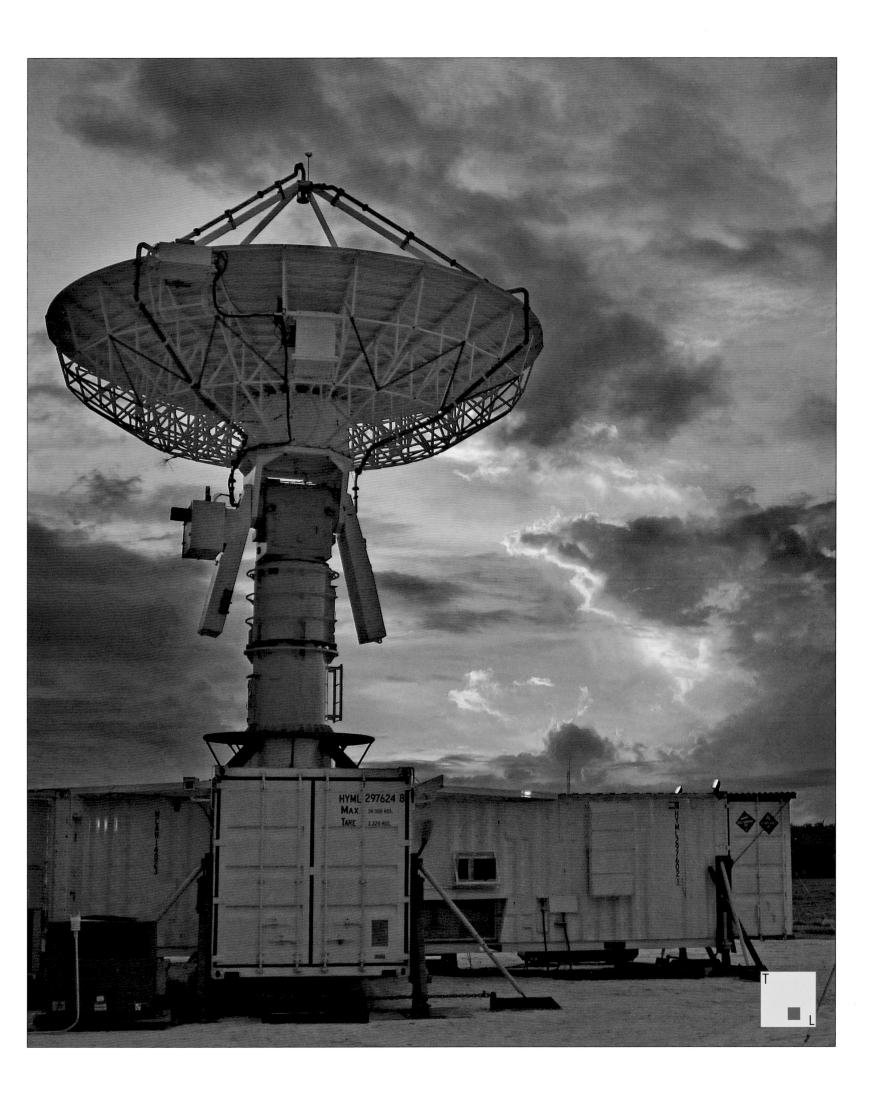

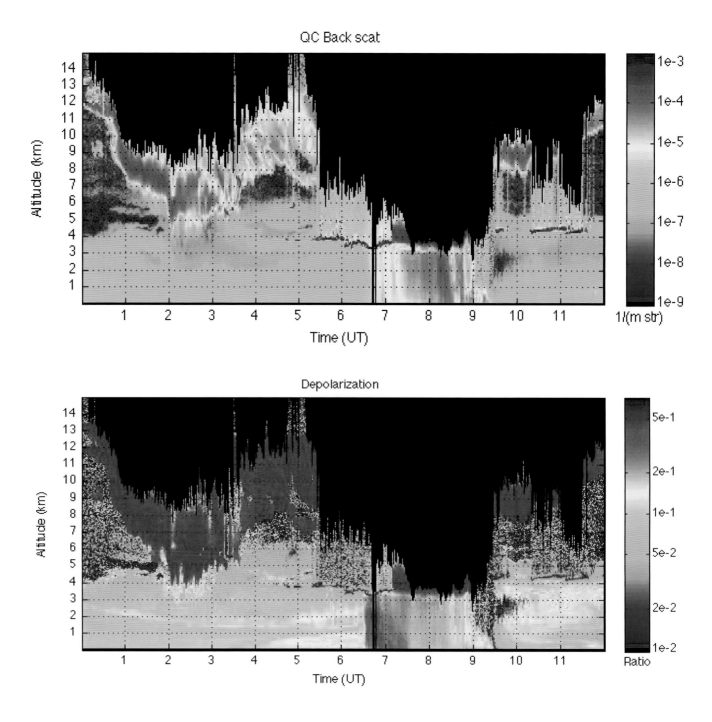

QC Back scat

Depolarization

Lidar backscatter measurements are shown in the first two panels. Atmospheric particle types that were retrieved from lidar measurements are shown in the third panel.

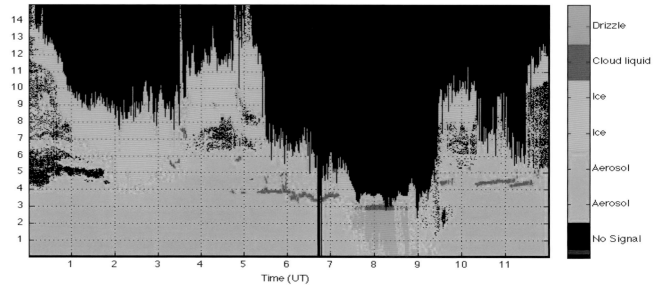

Classification

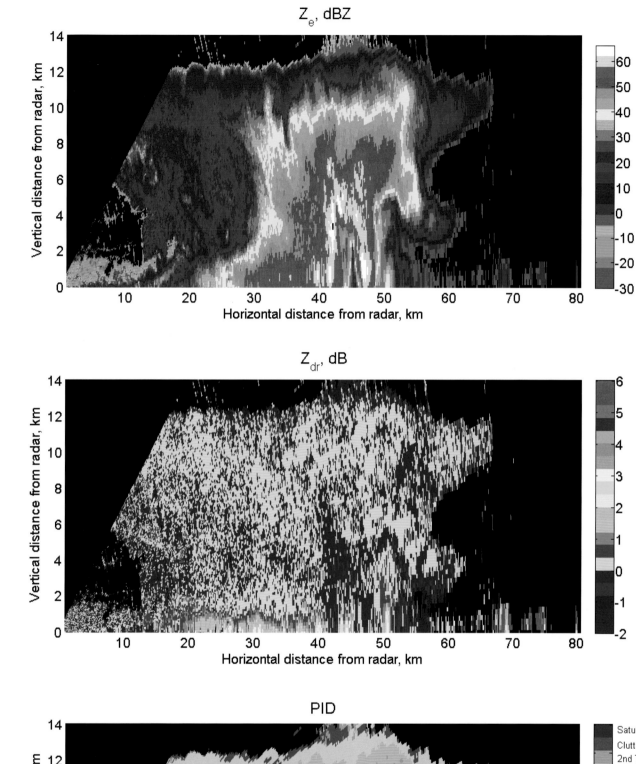

Z_e, dBZ

Vertical distance from radar, km

Horizontal distance from radar, km

60
50
40
30
20
10
0
-10
-20
-30

A vertical cross section of a thunderstorm observed by S-band microwave radar. Reflectivity and differential reflectivity measurements in the first two panels are used for identifying various precipitation types as well as insects.

Z_{dr}, dB

Vertical distance from radar, km

Horizontal distance from radar, km

6
5
4
3
2
1
0
-1
-2

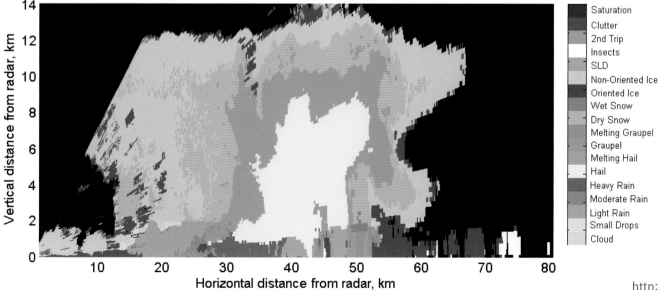

PID

Vertical distance from radar, km

Horizontal distance from radar, km

Saturation
Clutter
2nd Trip
Insects
SLD
Non-Oriented Ice
Oriented Ice
Wet Snow
Dry Snow
Melting Graupel
Graupel
Melting Hail
Hail
Heavy Rain
Moderate Rain
Light Rain
Small Drops
Cloud

LIDAR – Light Detection and Ranging instrument
http://youtu.be/E2w0B8VmwWQ

HYDROLOGY, A SCIENCE OF WATER FLOW AT MANY SCALES

Alberto Guadagnini, Erwin Zehe, Hubert H.G. Savenije, Alison D. Reeves
Hydrology and Earth System Sciences (HESS)

Water is the lifeblood of Earth. All living organisms need water to thrive and the evolution of our planet is dependent on it. Our landscapes are formed and shaped by water, our society depends on water, be it as rainfall, water in rivers, lakes and subsurface water. We suffer when there is too little or too much of it. Droughts and floods wreak havoc across our cultural landscapes, shaping our heritage. Water contamination is an additional concern for modern society. Water borne diseases are among humankind's worst enemies.

Given its importance to the evolution of society, it is not surprising that scientists have always had a keen interest in continuously increasing the level of knowledge on how water flows on Earth. It is a fascinating subject, comprising phenomena ranging from the visually compelling intricate patterns of water flowing over boulders to the devastating powers of floods forcing their way through natural and built environments. The flow paths are interconnected, incredibly heterogeneous and dynamic, the transport of water and solutes in the soil and the bedrock, overland water flow over complex topography, the invisible vapour fluxes associated with evaporation in terrestrial ecosystems, the mixing and spreading of chemicals in lakes, the intrusion of saline waters in estuaries and aquifers are just a few examples of how the flow of water links diverse parts of the Earth system.

Fascinating as the subject of hydrology is, studying it is fraught with difficulties because the physical, chemical and biological processes associated with the flow of water take place at a wide range of space and time scales, from individual pores in the soils to geological formations and entire river basins at the continental scale, from seconds to millennia. Because of interactions, pore scale processes have strong effects on the way solutes migrating in water spread over large water bodies and in the atmosphere. Subtle and complex reaction patterns due to micro-scale mixing may eventually limit the availability of nutrients, hydrocarbons and other chemical compounds. Preferential flows, where most of the fluid flows across a small part of the system on the surface and in the subsurface, tend to control the large scale fluxes. Similarly, most of the mass, energy and momentum is transported during short, extreme events, including floods, landslides, sediment flows, and pulses of contaminants. These events are often triggered when thresholds are exceeded, such as rain exceeding the infiltration capacity of a soil or stream velocities exceeding the resistance to motion of sediments in a stream.

Some hydrological processes, such as the percolation of water and contaminants through a soil column, can conveniently be studied in the laboratory. At the field scale quite different processes may become dominant, such as flow in large-scale fractures (not examined in the laboratory). This causes a scale problem. How do we transpose what we have learned at the small scale to larger scales? The main difficulty lies in the fact that the water fluxes are governed not so much by the properties of the fluid, but rather by the structure of the medium through which it flows. This structure has emerged as a result of the interaction between the water and the medium. Hence, it reflects the legacy of the flow characteristics of the water.

There are also a range of measurement methods at larger scales, pumping and tracer tests for inferring aquifer characteristics, a whole set of remote sensing instruments to infer soil moisture, snow cover and vegetation on the land surface, and flow measurements of streams draining catchments of hectares to millions of square kilometres. Here a critical challenge is to properly interpret the data in light of the small-scale processes.

Hydrological schools of thought have linked scales in different ways. One approach is to derive effective parameters that represent the averaged (large scale) flows using small-scale equations. Examples are hydraulic conductivities that average the system behaviour at sub-Darcy scales, or transmissivities, which arise in two-dimensional average settings. Another example are the dispersion coefficients governing transport processes that can be related to both the underlying flow field and the physical and chemical heterogeneity of an aquifer. An alternative is to build equations that explicitly capture the large scales. These are often including a series of additional terms to account for scale effects. There are also a range of stochastic methods that acknowledge the lack of local information at all scales and capture the flow and transport processes by stochastic differential equations. Their advantage is the possibility to explicitly represent uncertainty, which helps when assessing the level of confidence one can have in predictions.

Flood of the Swat River in Pakistan. Extreme flood events occurring in otherwise minor streams are examples of water flow behaving completely differently above and below a threshold. *Image taken from video referred to on page 73.*

Outflow of the karstic Kläfferquelle spring in Austria. Note the intricate patterns created by the interaction of water flow and topographic structure.

Underlying all these up-scaling methods is the idea that the detailed understanding of small scale processes can be used, in some way, to predict processes at larger scales. This may not always be the case, in particular when emergent behaviour occurs. The latter comprises, for example, the meandering of streams, the organization of sedimentary deposits of aquifers into distinct layers, the organization of vegetation in the landscape into patches, or the patterns of soil in the landscape imposed by the interaction with plant roots. Such emergent behaviour is often controlled by large-scale concepts, such as the optimality of energy or entropy states. This is an alternative, more holistic viewpoint to the up-scaling concepts. The jury is still out on what ideas work best in what context.

Even as remarkable advancements have been made in understanding the flow of water and the transport of substances by water in the Earth system, much is left to learn. At the heart of future progress will be our ability to observe phenomena at the smallest scales and use this information to enhance our knowledge of the fundamental mechanisms acting at the larger scales of rivers and aquifers, i.e., where they are most needed by society. Equally important will be our ability to understand the processes directly at the scale of entire landscapes, river basins and aquifers that are governed by holistic rules of optimality, and link them to the fine scale patterns that are so obvious to the eye. Identifying and describing all the complex patterns so typical of hydrological systems and the equations that describe the flow through these patterns may be the most exciting challenge for hydrology in the years to come.

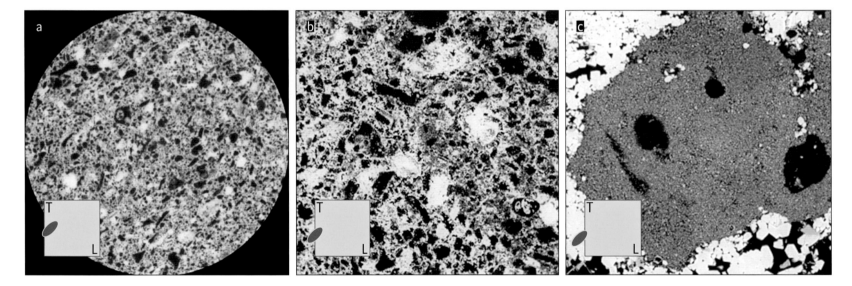

Cross sections of a Mallorca limestone sample X-Ray Micro Tomography image (a) and images of thin sections (b, c) mapped by an Environmental Scanning Electron Microscope at two different scales.

HESS Executive Editors

ERWIN ZEHE, *Karlsruhe Institute of Technology KIT (Germany); erwin.zehe@kit.edu*
ALBERTO GUADAGNINI, *Politecnico di Milano (Italy); alberto.guadagnini@polimi.it*
ALISON D. REEVES, *University of Dundee (UK); a.d.reeves@dundee.ac.uk*
HUBERT H.G. SAVENIJE, *Delft University of Technology (Netherlands); h.h.g.savenije@tudelft.nl*

Flood, Swat river, Pakistan
http://youtu.be/FPdxEU75UYk

Dead Sea sinkholes. The formation of sinkholes is a stark example of the way hydrological processes interact over multiple spatial and temporal scales. Starting from mixing and reaction of waters with different chemistry within the pores, the process slowly evolves, possibly interacting with vegetation, and dissolving the porous matrix. This leads to the formation of a network of cavities which at some point can cause sudden collapse of the ground.

SCALES AND SCALING IN NATURAL HAZARDS

Bruce D. Malamud
Natural Hazards and Earth System Sciences (NHESS)

The term natural hazard includes a wide range of diverse physical phenomena, both here on Earth and extra-terrestrial. We can broadly think of natural hazards as falling in six distinct hazard groups: geophysical (e. g., earthquakes, tsunamis, landslides, volcanic eruptions); hydrological (e. g., floods, droughts); shallow earth processes (e. g., ground collapse); atmospheric (e. g., cyclones, tornadoes, hail, lightning); and biophysical (e. g., wildfires).

When we think of natural hazard events (i.e., natural disasters), we often think of sudden and large events with devastating impacts that occur in a short amount of time, such as a large wildfire, earthquake or snow avalanche. These 'large ones' are what gain the attention of the press, decision makers, and the public. But, natural hazard events are not just the rare large ones, but also the medium and small ones.

Take for example the wildfire on the left. This wildfire on 8 August 2013 burned more than 24 square kilometres near Banning, about 140 kilometres east of Los Angeles. A thousand firefighters were involved in battling this blaze and about 1500 people had to flee from the flames. In this example, the wildfire is large. But, wildfires occur at many different scales, both in space and time, with the frequency, intensity, pattern and severity of their impact varying over many orders of magnitude, and dependent on such items as vegetation type, season, wind, local relief, and human influences. In space, an individual wildfire can vary from fractions of a hectare (100 hectares is 1 square kilometre) to thousands of square kilometres, more than six orders of magnitude in area. In time, the lifetime of an individual wildfire can range from seconds to many weeks (i.e., millions of seconds), again more than six orders of magnitude. Thus wildfires range over many orders of magnitude both in their spatial extent and their temporal duration. This is not uncommon for many natural hazards, and are the results of the complex processes (including both environmental and human) by which each hazard event is created.

In addition to the many scales over which wildfires occur (true for many different natural hazards), one can also consider their scaling. Scaling has varying meaning to different groups of scientists, but here, we will take it as comparing some attribute at small scales to the same attribute at large scales (or vice-versa), either in time or space, and observing how the scaling changes. For example, of key interest to many decision makers and the public is how large an event will occur in a given period of time. To do this, we examine the probability of an event of a given size occurring — how many large, medium, and small events occurred in the past using historical records, the frequency-size distribution (i.e., the probability that a given size event will occur plotted against the size). We can then compare the attribute of 'how many' at small scales to medium scales to large scales.

The question for people who study natural hazards, then, is what kind of probability distribution do these events follow? Unfortunately there is no simple answer to this question, not least because extreme events are very rare. Let us examine this in the context of wildfires. For wildfires, researchers have found that there are very few large wildfires, more medium ones, and lots and lots of small ones. What is of key importance though, is that the shape of the resultant probability distribution that describe wildfire frequency-area relationships (i. e., the scaling relationships) have increasingly been found to be heavy-tailed in terms of their scaling. To illustrate the word 'heavy-tailed', consider a Gaussian, or normal, distribution that is symmetrical about a central peak. The values of the distribution fall away on both sides of the peak, and the 'tails' are those values to the far left and far right. In the Gaussian distribution these tails are exponentials. In contrast, a distribution with a heavy tail decays much more slowly, such as a power-law. Wildfires over many orders of magnitude of area follow inverse power-law distributions, which when plotted on logarithmic axes are inverse straight lines. The scaling then of a frequency-size distribution that is power-law is said to be 'scale invariant', as the ratio of the log of the number of large to medium, and medium to small, and small to very small, is the same — there is no inherent scale in the underlying frequency-size distribution — it is scale invariant.

It is not just wildfires that exhibit these scale invariant distributions. Many other hazards have for part or all of their range of sizes a similar frequency-size distribution. Take for example the snow avalanche from Pakistan shown in the image overleaf. Snow avalanches can be small and they can be large, with many more small snow avalanches occurring than large ones. Or, on pages 78 and 79 we see the tracks of all known Atlantic tropical cyclones from 1851 to 2012,

A helicopter fighting the 8 August 2013 fire near Banning, California.

with colours representing the wind speeds. If we examine the number of cyclones with high vs. medium vs. low wind speeds, the distribution is again heavy-tailed, with many more wind speeds that are 'low' vs. those that are very 'high'. Not surprisingly from what we know about nature, the largest events (high wind speeds, or very strong hurricanes) are much rarer than the much more frequent smaller events (lower wind speeds, or the tropical storms/depressions).

Where do these heavy-tailed, or scale-invariant distributions come from? Scale invariance appears to underlie many natural systems. For example, when we look at a photograph of a mountain range, a drainage network or ripples of sand in a river, we need some familiar object in the photograph to indicate the scale because the shapes appear approximately similar at each scale. Examine again the wildfire on the previous page — if one looks at the wildfire itself, one has no idea of the scale without some human element in the photo to give us an idea of its size. The same with the photograph of the snow avalanche to the right — without a human scale in the photo, we have no idea of how wide, tall or how much volume it takes up. In many images shown in this book, it is similar, we have no way of telling how wide the structures are if we have no scale. This idea of scale-invariance in nature can be extended to the shape of probability distributions for natural hazards. These statistical distributions have no inherent scale to them, no characteristic 'average' value.

Why are these distributions important? It is the extreme tails that give an idea as to how often, on average, the largest events might occur. The decisions that might be made based on the shape of these tails are numerous: how high a bridge should be built, how deep electricity pylons need to be anchored, and so on. With very little historical data to play with, however, we sometimes have to make our best guess as to what the tails actually look like. If we pick a distribution where the tail decays quickly, then large events are very rare. In a distribution with heavy tails — where there is no inherent scale, on the other hand, large events are much more likely. The shape of the tail for natural hazard probability distribution sizes is our first step to assessing risk — heavy-tailed or power-law distributions are much more conservative in their estimation of risk compared to Gaussian (exponential) tails — so the implications for humans on which scaling distribution is chosen, can literally be thousands of millions of euros in building costs or mitigation measures as we decide whether a given size event (or larger) has a 1 in 100 chance of occurring next year or 1 in 1000 chance of occurring.

Where is the natural hazard community headed in terms of issues of scales and scaling? The first is increasing the awareness of others outside the community that natural hazard events are not 'just' large ones, but also occur at medium, and small and very small spatial and temporal scales. This awareness is important for the general public, that they not just think about 'the very large' events, but at all scales, so they can make appropriate plans. Second, is to ensure that the computational models that we have include in their theoretical designs the observations that we have on scaling of hazards. Not all models currently do include what we have observed in nature, but increasingly the community is becoming aware of the many issues around scaling of hazards.

NHESS Executive Editors

BRUCE D. MALAMUD, *King's College London (UK); bruce.malamud@kcl.ac.uk*
FAUSTO GUZZETTI, *CNR Perugia (Italy); f.guzzetti@irpi.cnr.it*
STEFANO TINTI, *University of Bologna (Italy); stefano.tinti@unibo.it*
UWE ULBRICH, *Freie Universität Berlin (Germany); ulbrich@met.fu-berlin.de*

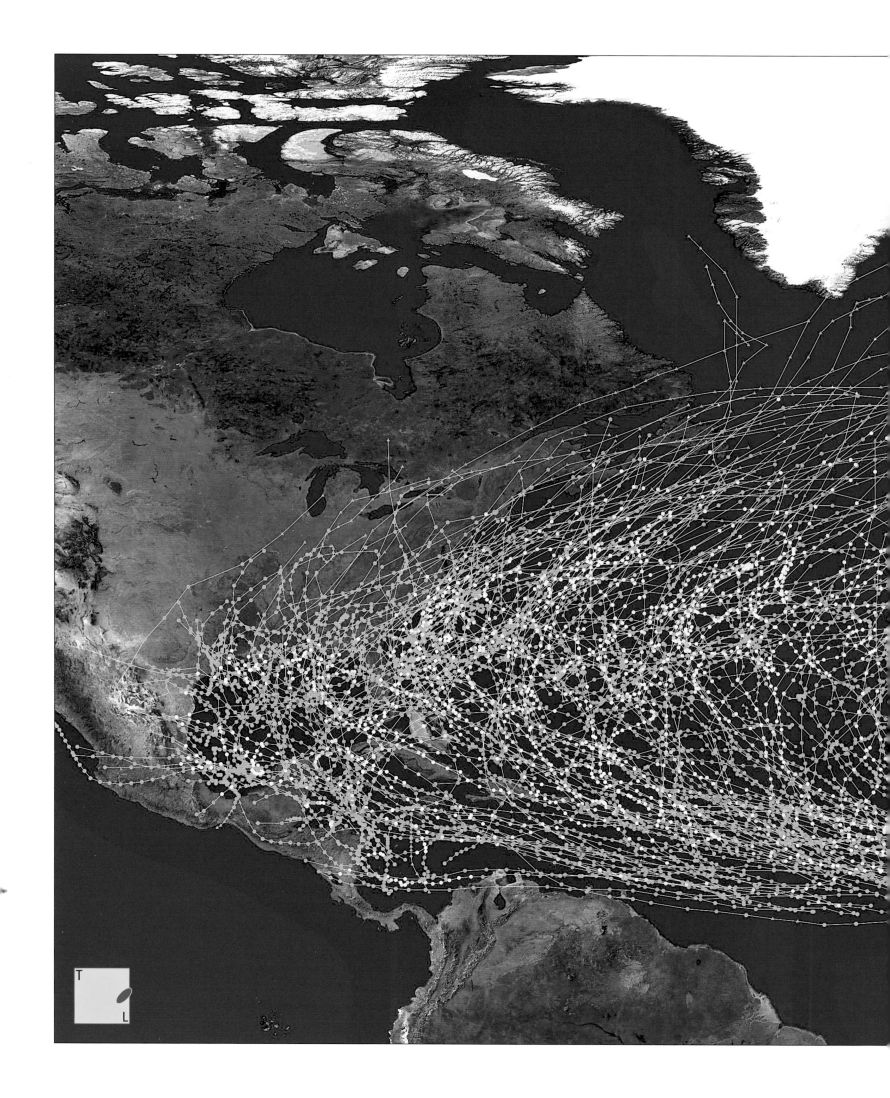

Gansu China Active Landslide
13 September 2012
http://vimeo.com/50837463

Tracks of all known Atlantic tropical cyclones, 1851 to 2012. Points show storm locations at six-hourly intervals. Colours represent the Saffir-Simpson Hurricane Wind Scale (see legend) with blues tropical depressions (TD) and tropical storms (TS), light yellows hurricane class 1 (winds 119–153 km/h), and red hurricanes class 5 (winds ≥252 km/h).

TD TS 1 2 3 4 5

BEAUTIFUL GEOMETRIES UNDERLYING OCEAN NONLINEAR PROCESSES

Ana María Mancho, Jezabel Curbelo, Stephen Wiggins,
Víctor José Garcia-Garrido, Carolina Mendoza
Nonlinear Processes in Geophysics (NPG)

Finding order in the apparent chaos that seems to govern ocean motions is a formidable task which has drawn the attention of scientists and oceanographers all over the world for the last decades. The endeavour of describing how heat, salt, carbon dioxide and other biogeochemical tracers are transported in the ocean has become a global challenge, and its understanding is of vital importance for predicting and assessing their impact on global climate change or the distribution of natural marine resources.

The main agents that drive ocean circulation at different scales are large-scale ocean currents such as the Gulf Stream, Kuroshio, Antarctic Circumpolar Current; mesoscale eddy structures such as the Agulhas rings, or sub-mesoscale eddies found near coasts. Nowadays, these structures can be directly observed by satellite altimeters, high-frequency coastal radars or through in situ devices. While in the past, navigators such as Juan Ponce de León noticed their presence on expeditions.

While strong currents are thought to be responsible for most of the major transport processes, eddies determine much of the mixing in the ocean. However, a common aspect of both features is that they define regions where fluid parcels have difficulty crossing, which are known in the literature as transport barriers. These water barriers form boundaries between fluids with different physical properties and give rise to an organizational structure in the overall picture of ocean flows. Eddy-induced transport has been historically underestimated, but increasingly it is being recognized as a key contributor to ocean transport processes. Indeed, since vortices are robust, long-lived structures that may persist for periods lasting from months to years, when water gets eventually trapped within the eddy's core, the fluid inside will travel hundreds to thousands of kilometres within the mesoscale structure and preserve its biogeochemical properties for a long time.

The ocean motion follows a fully nonlinear dynamics in which turbulence is an essential feature, enabling interactions between motions on different spatial scales and playing a key role in ocean transport and mixing. Efforts to understand and parameterize turbulent mixing have been a research focus for many years, and continue to be essential towards understanding and predicting the evolution of the Earth's oceans.

In this context, dynamical systems theory has provided a framework for describing messy paths of float trajectories in ocean flows. Behind their apparent disorder, a subtle and sophisticated order is revealed through geometrical structures over many scales. The underlying fabric uncovered by dynamical systems tools has been considered as the skeleton of turbulence. These structures, which organize trajectories schematically into distinct regions corresponding to qualitatively different types of trajectories, are responsible for the transport and mixing processes that govern the ocean in these areas. The boundaries, or barriers, between these regions are mathematically realized as objects called manifolds. The figure on the left makes visible this geometrical representation for the Gulf Stream on the 14 March 2004 over an area extending 1100 km in length. The image is produced using altimeter data sets distributed by AVISO (http://www.aviso.oceanobs.com/duacs/). This representation is achieved using Lagrangian Descriptors, a mathematical tool which highlights invariant manifolds in flows with a general time dependence.

A magnification of the area indicated by the rectangle is shown on the next page right. An intricate geometrical pattern, a chaotic tangle, is observed at a lower scale, in which the filamentary structures are less than one kilometre in length. The skeleton displayed in these figures represents time dependent dynamical barriers along which passive ocean tracers are attracted either backwards or forwards in time.

The figures summarize particle histories over a time interval of 140 days: 70 days after the 14 March 2004 and 70 days before this date. Should this time interval be reduced then the complexity of the figure is also reduced.

The six small figures show a sequence of images summarizing particle histories. They show a small region of the Gulf Stream with a different colour scale and forward and backward time intervals of 15, 30 and 50 days, respectively. Filamentary patterns are related to the mixing of passive scalars during the time interval over which the figures are calculated. The bottom panels show the forward evolution of two particle sets, red and blue, according to the Aref

Geometrical representation for the Gulf Stream on 14 March 2004 as revealed by
the Lagrangian Descriptor function. The image covers an area of 1100 by 600 km.

A Voyage Through Scales **81**

Magnification of Gulf Stream from previous page (270 by 150 km)

Below
Sequence of images summarizing particle histories
in the Gulf Stream with time intervals of 15, 30 and
50 days (top) and Aref blinking vortex maps (bottom)

blinking vortex map (a conceptualisation of flow patterns related to a pair of vortices, that switch on and off alternately). After a short flow period, particles are not mixed, their evolution does not show complicated patterns. However, at longer times this structure becomes more and more labyrinthine, showing that attracting filaments and particles are intermingled.

The structures shown in the figures are more than just beautiful mathematical objects obtained from the nonlinear ocean motion: they are objects truly present in nature. The image of sea ice off the east coast of Greenland on the following pages shows structures strikingly similar to those displayed in the analysis figures. The swirling structures marked out by the sea ice are visualizations of the unstable manifolds

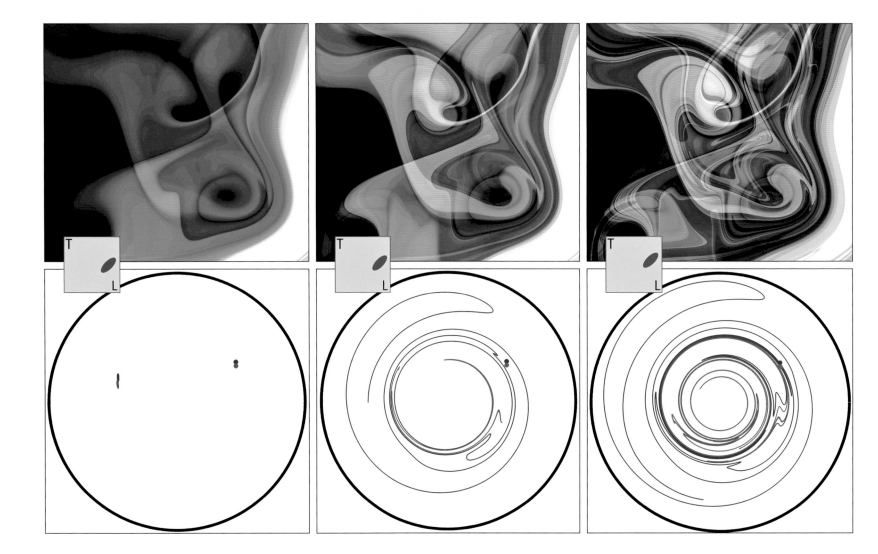

NPG Executive Editors

ROGER GRIMSHAW, *Loughborough University (UK); r.h.j.grimshaw@lboro.ac.uk*
JÜRGEN KURTHS, *Potsdam Institute for Climate Impact Research (Germany); juergen.kurths@pik-potsdam.de*
ANA M. MANCHO, *ICMAT, Madrid (Spain); a.m.mancho@icmat.es*
DANIEL SCHERTZER, *Université Paris-Est (France); daniel.schertzer@enpc.fr*
OLIVIER TALAGRAND, *CNRS Paris (France); talagrand@lmd.ens.fr*

Nonlinear Processes
in Geophysics
http://youtu.be/P-f4k-cjk_8

Sea ice off the east coast of
Greenland taken by NASA's Aqua
satellite on 17 October 2012.

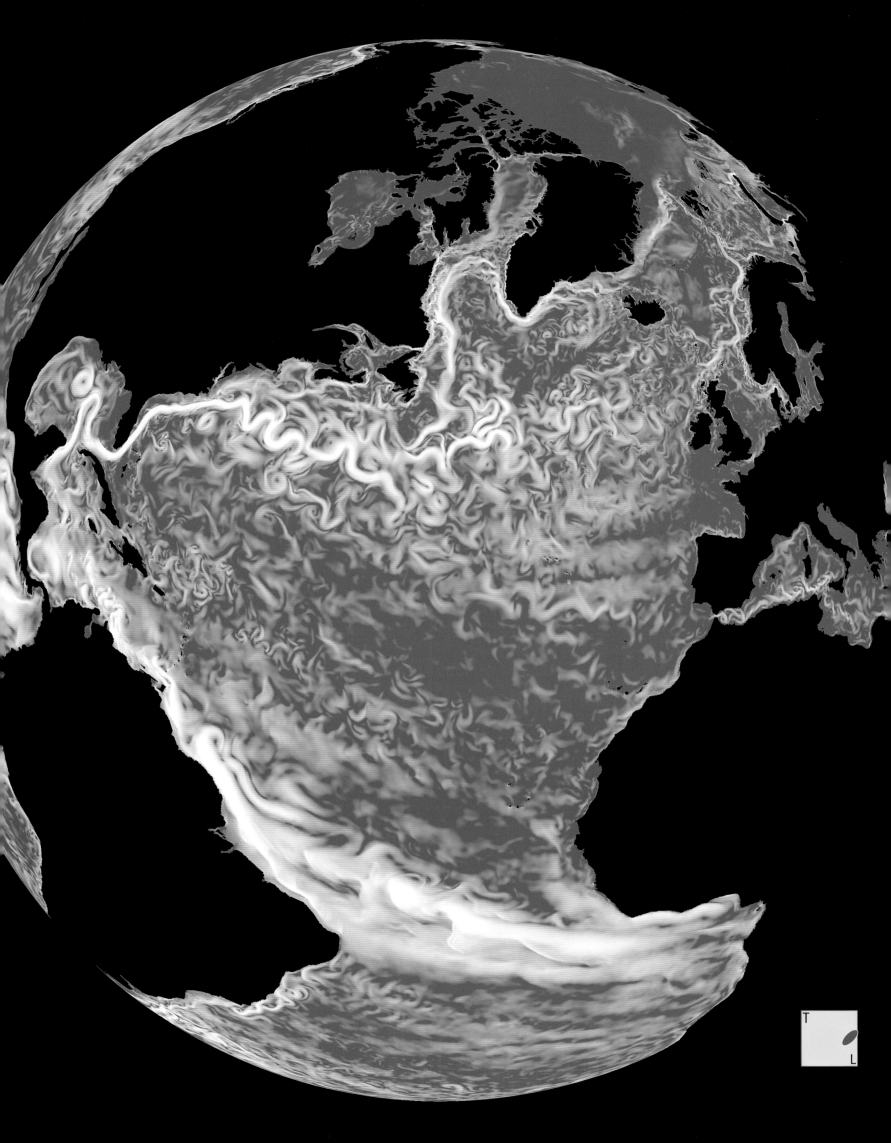

**Big whirls have little whirls that feed on their velocity,
and little whirls have lesser whirls and so on to viscosity.**
Lewis Fry Richardson

The seas cover more than two-thirds of the Earth's surface. They hold and transport heat and carbon, so affecting weather and climate; presently oceans continue to take up more than a quarter of anthropogenic carbon dioxide emissions. The seas are an important food source. Marine life has long fascinated mankind. Waves on the ocean surface, tides and locally strong currents have posed challenges to sailors for milliennia.

Modern society has many reasons for needing to know about and understand the oceans. These reasons include the design and legislation for expected climate and conditions for offshore operations (fishery, coastal developments and defence). Forecasts (early "weather" warnings for safe operations, coastal defence, Harmful Algal Blooms, response to spills, search and rescue, navigation and recreation) require advanced ocean-atmosphere coupling in computer models. Discharge consents impact on water quality (toxins and lack of oxygen). Sustainable use of resources (minerals, renewable energy) entails their management and knowledge of seafloor and sub-seabed processes to detect natural resources and geo-hazards, and of habitats for living marine resources including fish. To detect the effects of climate change such as "ocean acidification"; where and how heat and carbon penetrate into the deeper ocean affecting the Earth's climate. Ocean warming is estimated to account for 90 percent of the energy accumulation from global warming between 1971 and 2010. Warming oceans expand, an important part of sea level rise.

The EU Marine Strategy Framework Directive, for example, aims to "protect and preserve the marine environment" and "ensure that there are no significant impacts on or risks to marine biodiversity, marine ecosystems, human health or legitimate uses of the sea". These interests, as well as scientific curiosity about a large part of our planet which is still relatively unexplored, give us strong reasons for the study of ocean science. The scope includes: ocean physics (temperature and salinity distribution; currents, eddies and circulation; sea level, tides, waves on the surface and in the interior; turbulence and mixing); ocean chemistry (sources, transport and cycling of various chemical substances including oxygen, nitrogen, carbon, phosphorous and iron, affecting climate and ecosystems); air-sea interactions (exchanges of heat, momentum and gases between sea and air; sea-ice); biological oceanography (plant growth, productivity and successive "higher" levels of marine organisms, ecosystem dynamics, biological diversity and ecosystem function in the deep sea); the sea floor (plate tectonics, geology; sediment processes and bed-forms from ripples through sand banks to continental shelves and slopes, ocean basins; from hydrothermal vents to mid-ocean ridges); and paleoceanography (the history of the oceans in the geologic past).

The coverage is worldwide, from the surf zone at the shore to the deep ocean. Moreover, the shallower seas over continental shelves adjacent to coasts have different characteristics and strong impacts at the coast. The time-scales of impacts range from days to decades and spatial scales from local to global. The oceans are therefore grossly under-sampled and different approaches are needed to build a picture: instrument development; innovating technology aiming for continuous, adaptive, autonomous ocean measurement by marine sensors and platforms; remote sensing; laboratory and theoretical studies; computer models (some now with data assimilation) for physical, chemical, biological and biochemical aspects. "Platforms" include satellites and aircraft, moorings, drifting buoys and floats (many do profiles), ships especially for detailed transects and surface measurement of some variables while under way, coastal stations and radars. By these varied means we seek to understand fundamental biological, chemical and physical processes and to determine how patterns and processes on smaller scales drive larger-scale features and vice versa.

Spatial scales related to ocean processes range from molecular and turbulent (microns to millimetres) to planetary (thousands of kilometres). Controlling factors include viscosity (friction) for turbulence and thin layers against boundaries, and grain size for sediment ripples. Water depth controls larger-scale bed forms such as sand waves and banks with length scales of metres to kilometres. Density differences depend on heating/cooling rates, specific heat and density dependence on temperature and salinity. Gravity, earth's rotation and density differences through

Ocean circulation graphic. This illustrates ocean-basin-scale circulation (the largest "whirl") and eddies (relatively "little whirls"), especially feeding off the strongest currents.

depth combine to form a typical length scale of kilometres to tens of kilometres for eddies and various internal deformations of the density distribution. Sea-level variations (e.g. tides) have a scale of hundreds to thousands of kilometres through a combination of gravity, earth's rotation and water depth. Topography from bed-forms through canyons to ocean basins directly imposes varied depth and lateral scales from local to global. Basin-wide ocean circulation is an example of the latter.

These physical aspects also control dispersion of chemical constituents. Some constituents may have very local sources (e.g. rivers, factory discharges). By contrast, atmospheric inputs may be widespread, but may be concentrated near the surface initially. Most of the salt in the ocean has been there for thousands of years and has had time to mix fairly evenly throughout ocean basins, except near river inflows.

The marine ecosystem shows a range of sizes for biota, typically increasing "up" the food chain. Thus basic plant life may be sub-millimetre scale; it depends on light from above and on nutrients, so such "primary production" often "blooms" for a week or so near the surface in spring. Once near-surface nutrients are depleted, it may be concentrated at the base of an upper warmer layer in summer. Some animals feeding on this may migrate 100 m or so up and down, diving daily to avoid predators. Whales may exceed 10 m length and travel thousands of kilometres seasonally. Corals are individually small but colonise to form reefs that may extend for hundreds of kilometres. On scales of thousands of kilometres, marine ecosystems have been classified into about 40 "provinces".

Time scales similarly have a broad range, from fractions of a second (millimetre-scale turbulence) to millennia (ocean property distributions) and millions of years for ocean basin evolution. Typical periods (1—20 seconds) of wind-driven surface waves derive from the ratio of wind speed to gravitational acceleration. Internal deformations typically have periods between some minutes (set by gravity and the rate of density increase with depth) and half a day to a few days (set by the Earth's rotation and depending on latitude). Weather-driven flows and changes of surface elevation ("storm surges") adopt the atmospheric weather time-scale of a day or two. Tidal oscillations relate to the length of solar and lunar day. Temperature, near-shore salinity and hence density-related currents typically vary through the year with seasonal heating, rainfall and light, as does marine life. Slow waves in the ocean interior may take years to cross an ocean basin. (Re-) adjustment of heat in the upper hundreds of metres (e.g. to a change in received surface heating) is on a time-scale of decades, as are circulation time around a large ocean basin and possibly "regime" shifts in ecosystems. "Climate" is usually taken to refer to statistics for at least a few decades (often 30 years) but even over such a long period there is evidence of reversible variability, e.g. the "Atlantic Multi-decadal Oscillation". Overturning circulation (eg. sinking in the Nordic seas, due to winter cooling, and complementary upward transport further south in the Atlantic and in other oceans) takes many centuries.

OS Executive Editors

WILLIAM JENKINS, *Woods Hole Oceanographic Institution, Massachusetts (USA); wjenkins@whoi.edu*
ERIC J.M. DELHEZ, *University of Liège (Belgium); e.delhez@ulg.ac.be*
JOHN M. HUTHNANCE, *NERC National Oceanography Centre, Liverpool (UK); jmh@noc.ac.uk*

Individual coral components on millimetre scales colonise to form clusters on metre scales and whole reefs which may extend to hundreds of kilometres. Papua New Guinea.

The NEMO global ocean model
http://youtu.be/I8ru1YvXU04

Sand banks and sand waves off Beaumaris, Anglesey, Wales showing scales from metres to hundreds of metres. These features are exposed at low tide; at higher stages of the tide they are covered and evolved over days to decades by strong tidal currents. Bedforms range from ripples to ocean basins with mid-ocean ridges. Generally, the larger the length scale, the longer the evolution time. Ripples typically come and go each tide, sand waves perhaps come and go with the seasons or big storms, and sandbanks such as the largest features here may take decades or centuries to evolve.

SOIL: A JOURNEY THROUGH TIME AND SPACE

John N. Quinton, Jorge Mataix-Solera, Eric C. Brevik,
Artemi Cerdá, Lily Pereg, Johan Six, Kristof van Oost
SOIL

Soil is the life support system of our planet. It helps make our air breathable, cleans the water we drink and supports production of the food we eat. This life support system relies upon processes that operate at spatial scales from less than a micron to over hundreds of metres or more, and over timescales from seconds to millennia.

The smallest soil particles are nanometres across. It is at this scale that we find the engine room of the soil where chemical compounds are transformed between gas and liquid phases and where material containing carbon is digested by microorganisms, who then release carbon dioxide. Just like animals and plants living on the soil surface, most microorganisms need air and water to survive. At the nanometre scale the soil atmosphere has a carbon dioxide concentration that is much greater than the air we breathe, and where soil pores are filled with water there may be little oxygen. In this anaerobic world we find microorganisms that are specially adapted, relying on other compounds for respiration. Processes occur quickly here — microbial life cycles may take a matter of minutes and in that time they perform many vital biochemical processes for the soil.

One of the reasons that many chemical transformations take place at the nanometre scale is the enormous surface area of these tiny particles — just a gram of soil may have a surface area equalling the size of many football pitches — providing plenty of opportunity for chemical reactions. The surfaces of the tiniest soil particles also carry an electric charge which repels some ions and attracts others. The ions held on these surfaces can exchange with each other in response to changes in their concentration in the soil solution; others diffuse into the mineral structure and are held more strongly, but can become exchangeable again if conditions change. This process is responsible for supplying plants with nutrients that are vital for life. Managing the availability of these nutrients is critically important for food production.

This miniature material is often bound together into soil aggregates, ranging in size from a few microns to several tens of millimetres in size, which organize the space within the soil into pores that are vital for transmitting water and air, as well as providing habitats for soil organisms. Often soil aggregates join together to form a complex arrangement of spaces and solid material, known as the soil's structure. It is this structure that farmers try to manage to produce a seedbed that will allow plant roots to penetrate so they can reach water and nutrients held within the soil. It is also incredibly important for storing and transmitting soil water — almost all the water supplied to plants is stored in the soil. However, soil structure can easily be damaged, either by heavy machinery or wind and rain: machinery can crush the soil, squashing its largest pores; raindrops break up soil aggregates into small particles that can block soil pores; and wind can remove soil and deposit it on other soil surfaces. These can all make it harder for water to percolate into the soil, reducing the soil's ability to absorb water, which can lead to erosion and flooding. Thus a change in soil structure at the scale of a few micrometres can lead to effects at much larger scales.

Dig a hole in the soil and you will eventually reach rock. In some places this will take you just a few minutes as some soils are only a few centimetres deep; however, find yourself in the tropics and you could be digging for several days as you try to reach the rock which may be several metres below the soil surface. It is clear that soils are not all the same and in fact they vary incredibly across landscapes. Even within a field, the bottom of hills can receive soil material from further upslope resulting in deeper, often more fertile soils, leaving shallower less productive soils on the steeper slopes. The scales of these changes are governed by the effects of climate, topography, geology, organisms and time.

Where climates change dramatically over larger scales, soils also change: a soil formed over granite in the humid tropics may be several metres deep and a deep red colour, while in cooler northern latitudes the soil formed over similar rock may be dominated by slowly decomposing organic matter and look almost black. But climate can also affect soils at much more local scales: the aspect of a hillside can affect how much sunlight it receives and how much water is available, influencing the species of vegetation that are able to grow.

So, climatic and soil changes affect the vegetation, which in turn influences the soil; making soil and vegetation a closely integrated system. Vegetation can influence soil properties at the micro scale by removing nutrients and

Soil layers of the Palouse region of eastern Washington are revealed in a deep road cut. More than ten layers of buried soils are visible to soil scientist John Reganold of Washington State University.

water, providing food and habitat for soil organisms and adding organic material to the soil as it decays. At larger scales vegetation is an important soil structure former, creating and blocking soil pores with roots and protecting the soil surface from destructive raindrops. The protection given to the soil surface by vegetation cover is important for controlling soil erosion — which is key for the sustainability of soils in the future — and influences the way in which soil material is redistributed in the landscape.

Soil erosion and soil redistribution is linked closely to topography, which itself operates at a range of scales. Many erosion processes are controlled by surface topography which may only change by a few tens of millimetres: on a very rough soil surface soil particles may only be redistributed at a very local scale, whereas on a smooth surface, such as a tilled seed bed, small rivulets of water can join together to produce flow that moves more soil material on a wider scale. When this happens, topographic variation over the scale of metres to tens of metres exerts a greater influence, concentrating runoff and erosion in valley bottoms and encouraging more soil to be deposited where slopes are less steep.

Many properties of soils are controlled by the rocks and deposits in which they form, from changes in the types and properties of soil minerals present, which influence soil processes at the nanometre scale, to changes between rocks that are more or less prone to mineral weathering, leading to soils with widely different depths and textures. For example, soils formed over limestone, which weathers slowly, are often shallow and contrast with deeper soils that form from soft easily weathered sandstones.

Soils are found across the surface of the globe, from the high artic to the deserts of the Sahara, and some soils have been around for a very long time. There have been soils on our planet for at least 3 billion years. While soils that old tend to be preserved in rocks as paleosols — just as dinosaurs were preserved as fossils — there are still some pretty ancient soils on the face of the earth. For example, some soils in Hawaii began to develop 4.1 million years ago. These soils are often deeply weathered and have had many of their nutrients leached out. We can see the effects of time on the development of a sequence of soils close to Alicante in Southern Spain. These soils are all from the same region, are exposed to a similar climate and have the same parent material (calcareous rocks) but have a very different appearance. This is because the soils of different ages have been exposed to weathering and other soil forming processes for different lengths of time, leading to variations in soil colour.

The wide range of spatial and temporal scales involved in the formation and function of the world's soils makes them a fascinating topic for scientists to study. Our species' absolute reliance on them for food, water, building sites, raw materials and as a home to many of our planet's living organisms makes understanding how time and space interact to influence the future of our soils essential.

SOIL Executive Editors

ERIC C. BREVIK, *Dickinson State University, North Dakota (USA); eric.brevik@dickinsonstate.edu*
JORGE MATAIX-SOLERA, *University Miguel Hernández, Alicante (Spain); jorge.mataix@umh.es*
LILY PEREG, *University of New England, Armidale (Australia); lperegge@une.edu.au*
JOHN QUINTON, *Lancaster University (UK); j.quinton@lancaster.ac.uk*
JOHAN SIX, *ETH-Zurich (Switzerland); jsix@ethz.ch*
KRISTOF VAN OOST, *Université catholique de Louvain (Belgium); Kristof.vanoost@uclouvain.be*

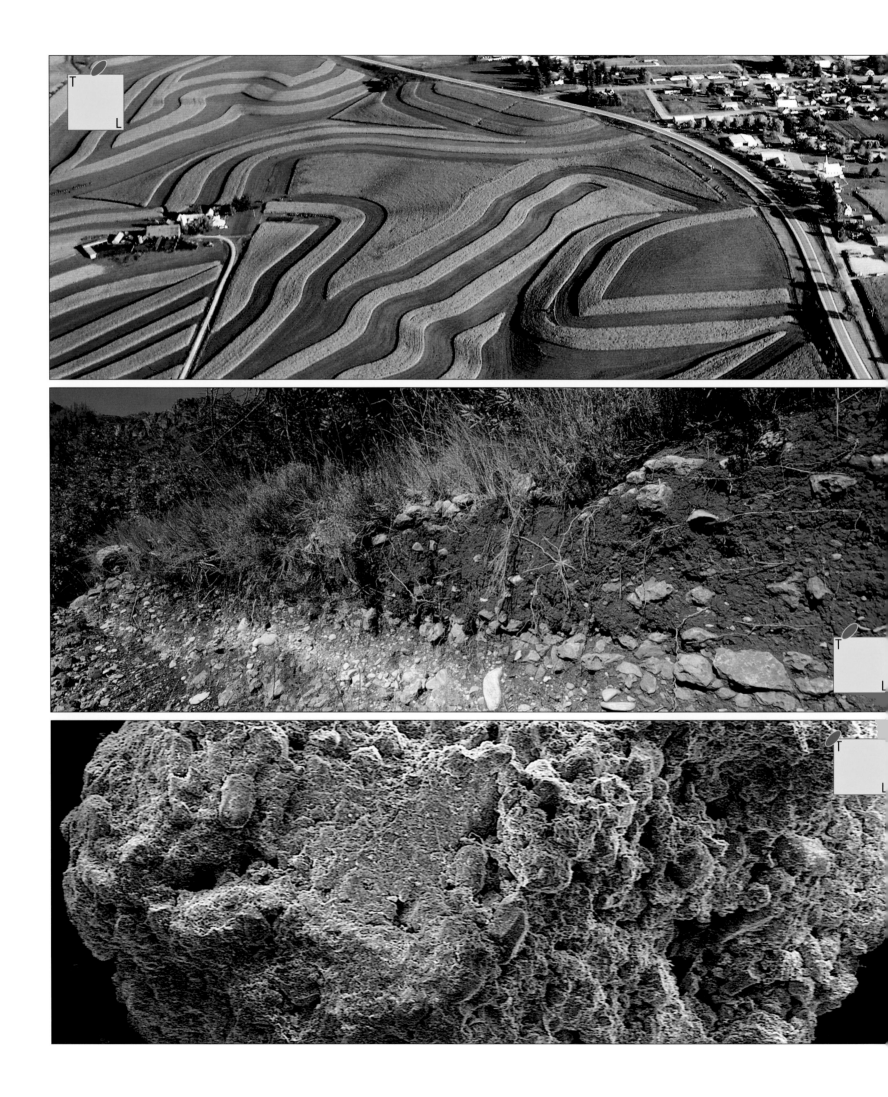

Viewing soil extracts under
the microscope
http://youtu.be/VuHznslr8aI

THE MOBILE SOLID EARTH: A QUESTION OF PERCEPTION

Charlotte M. Krawczyk
Solid Earth (SE)

The solid Earth is regarded by human beings as their foundation. On the Earth's surface we learn to walk and move around with steadily increasing distances. Even though only one third of the Earth's surface is solid, it continues below the oceans and thus constitutes the global anthroposphere.

The solid Earth reveals beautiful rocks and minerals at the small scale, and exposes spectacular folds and structures on a global scale. Assemblages of many very small particles can often produce a large structure, just as many sand grains can build a sand dune.

At the same time nothing is static within the solid Earth, just like the theories and hypotheses about the structure and processes inside its interior. All of the Earth's movements, changes, and deformation can occur on time scales that range from billions of years to fractions of a second. It can take years before, for instance, enough stress has built up to allow the rupture of material, such as earthquakes or artificially produced seismicity. Similarly, volcanic eruptions can happen spontaneously or become active after a long period of apparent dormancy.

The solid Earth is, hence, investigated and exploited in a myriad of ways. As such, it is equally important to understand the global processes of plate tectonics as well as it is to study small-scale local phenomena that, for instance, may trigger the decision how to utilize the subsurface, e.g. hydrocarbon exploration or infrastructure measures. This is aided by a large variety of different methods that address the specific properties of the subsurface.

When Alfred Wegener publically outlined his theory of continental drift for the first time in a talk ("New ideas about the development of large-scale structures of the Earth's crust based on geophysics", 1912), he had no view of the Earth from above and could not explain what actually drives the plate movement. Consequently the supporters of contraction theory maintained their authority. Exploration of the oceans, research on volcanism, earthquakes, fossils, and in more recent times space-borne and satellite techniques yielded, ca. 50 years later on, the relevant pieces of the puzzle to substantiate the theory of plate tectonics that revolutionized the geosciences. Today, the wander paths of the continents can be reconstructed for the last ca. one billion years.

We can measure and determine changes and variations of the Earth's surface using geomorphological and geodetic methods. Structural geology including the collection of samples and their analyses in the lab, the examination of soils, and sedimentology have all been used to describe phenomena seen at the surface. Their development from qualitative to quantitative research not only allowed work across different scales, but especially intensified the understanding of processes and bridged the gap between models that are derived from field work, lab work or numerical work. The simulation of tectonics and the generation of earthquakes in sandbox models are recent examples.

To look inside the Earth, indirect methods for exploration from the surface are needed, so-called non-invasive methods, as they have been used since the 1960s, e.g. seismic exploration of resources-bearing structures with vibrators as signal source. Since the 1980s, many national research programmes have used this reflection seismic technique to image, for the first time, the large-scale structure of the continents down to 100 km depth (e.g., COCORP, ECORS, BIRPS, DEKORP, Lithoprobe). In the course of this imaging, mountains roots, extinct and active subduction zones, and large sedimentary basins have been detected at depth and spatially surveyed.

The Earth's figure and gravity field are determined using geodetic and gravimetric observations and investigations. Supplementing marine and land-based systems with satellite systems has brought about enormous development in the last decades. It provided global spatial coverage as well as better temporal resolution and repeat cycles for sampling. The mass movement within the Earth's interior can be determined, for instance, by the recent, sequential satellite mission of CHAMP/GRACE/GOCE with an approximately five to ten times higher resolution than before. In addition to a better understanding of the geodynamic processes in the interior of the Earth using complimentary satellite techniques, they also revealed changes in the groundwater or of the Antarctic ice mass, thus yielding important perceptions and new constrains for climate and hydrological models.

This piece of chert containing a "Mermaid's Heart" was found in the Tamu Massif east of Japan where the biggest single shield volcano ever discovered on Earth had formed in Late Jurassic times.

Another geodynamic component that is generated in the deep Earth interior is the Earth's magnetic field, which is mainly generated by dynamo processes in the Earth's outer fluid core, and to a lesser degree by magnetized lithospheric rocks. This magnetic field protects us against solar wind particles and cosmic radiation, and depends on seasonal variations or magnetic storms. Once again, the now technologically-possible combination of land- and satellite-based measurements allows the study of phenomena like space weather conditions and the evolution of the geomagnetic field. Recent investigations suggest the additional influence of oceanic circulation.

Proper intervention into the solid Earth is enabled by boreholes that sample material in drill cores at up to 10 km depth. Here, especially the large, international research programmes, such as ICDP/IODP (International Continental Scientific Drilling Programme/International Ocean Discovery Programme) contribute essential knowledge on the fundamental structure of the Earth. Equally important are shallower research wells and small-scale campaigns that are undertaken to investigate the lithology of a region, to perform age dating, or to test the geotechnical properties of the Earth material.

In addition to disciplinary methodical research, it is always necessary to combine findings from the laboratory, the field, and numerical work. Only this will allow the calibration of phenomena indirectly-observed, to stabilize simulations, or to try to forecast and predict developments on geological and human timescales. Here, seismology (the science of earthquakes) serves as an example. Seismology not only covers seismic wave propagation that has led to knowledge about the Earth's interior structure but in combination with other disciplines, it has enabled better understanding of earthquake processes and attempts at early warning. The seismicity in a region is better understood if one includes historical records, and geological investigations after an earthquake yield information for source mechanism estimations; both of these are influenced by tectonics. Also important, stress indicators are provided by GPS-observations and in-situ borehole measurements.

Fascinating, and with huge potential for future developments, is the fact that the scale behavior of a method is not coercively proportional to the scale of the geo-process studied. For instance, the analysis of the smallest mineral components and the determination of their chemical signature and age provide evidence on large-scale processes such as how a mountain range developed, how far material has been transported, or which depositional environment existed during a certain epoch millions of years ago. In the same vein, material properties and their specific combination can ascertain whether a structure in the subsurface can be imaged at all, or how a geodynamic process may be influenced.

Solid Earth research is, thus, multidisciplinary research on the composition, structure and dynamics of the Earth that reaches from the surface to the deep interior at all spatial and temporal scales. It hosts, and ideally combines, geochemistry, geodesy, geodynamics, geomorphology, geophysics, mineral and rock physics, magnetism, palaeontology, petrology, planetary science, sedimentology, seismology, soil science, stratigraphy, structural geology, tectonophysics, and volcanology.

A pioneer of cross-disciplinary scientific thinking and research, that should be considered today more modern than ever, was Alexander von Humboldt (final opus "Cosmos. Concept of a physical description of the world", 1845-1862). Even though part of his many and specific observations were not correct in detail and had to be revised subsequently, von Humboldt's sense of the overall geoscience context and interrelationship of single disciplines, as well as of the intervention between instrumental and visual-graphical observation of Earth processes is possibly unsurpassable, as is this timeless citation: "The most dangerous of all world-views is the world-view of those people who have never looked at the world".

SE Executive Editors

FABRIZIO STORTI, *Università degli Studi di Parma (Italy); fabrizio.storti@unipr.it*
CHARLOTTE KRAWCZYK, *Leibniz Institute for Applied Geophysics, Hannover (Germany); lotte@liag-hannover.de*
PAOLO PAPALE, *Istituto Nazionale di Geofisica e Vulcanologia, Pisa (Italy); paolo.papale@ingv.it*

A **vibrator truck** in the desert south of Abu Dhabi (United Arab Emirates) serves as signal source for investigating the geological structure with reflection seismic surface measurements. Such campaigns provide detailed models of the subsurface that are essential for identifying drill locations for hydrocarbon exploration or research tasks.

The Early Earth and Plate Tectonics
http://youtu.be/QDqskltCixA
Earthquakes - Shock Waves Explained
http://youtu.be/_YLjlvJXhpg

Sedimentological and paleontological analyses of deep-sea cores obtained during IODP Expedition 339 (Mediterranean Outflow) are fundamental to understanding the history of the exchange of water masses between the Mediterranean Sea and the eastern Atlantic Ocean.

Brendan O'Neill
The Cryosphere (TC)

The cryosphere is dynamic on a wide range of scales. It includes all frozen parts of the Earth system, from ice crystals in the air, to seasonal snow, sea ice, lake and river ice, glaciers, ice sheets, and ground ice. The annual maximum extent of Earth's frozen realm occurs during the northern hemisphere's winter, when snow and ice blanket much of the continents and sea ice cover on the Arctic Ocean peaks. Over the past few decades, glaciers and Arctic sea ice have diminished at unprecedented rates. At longer time scales, the extent of the cryosphere fluctuates more dramatically between glacial and interglacial periods.

Phenomena in the cryosphere originate from one microscopic process: the phase change of water. At the smallest scale, freezing and melting are responsible for the diverse processes and landforms that shape cold regions.

Features of the cryosphere range from the microscopic to continental scale. In the atmosphere, water vapour in clouds allows ice crystals to grow. These crystals continue to enlarge as they collide with water droplets and each other, eventually forming snowflakes large enough to precipitate to the ground. The ensuing snow cover may melt seasonally, or accumulate over time, compacting and eventually forming glaciers, or at the grandest scale, ice sheets kilometres in thickness. The seasonal freezing of water in polar oceans forms sea ice that covers millions of square kilometres, while lakes and rivers in cold regions become choked with ice. The often-spectacular breakup of river ice in spring may cause severe flooding, one of many influences of the cryosphere on hydrologic regimes.

In the periglacial domain, the microscopic process of ice segregation can form millimetre-scale ice lenses in permafrost soils, and massive ice tens of metres thick. In addition, the contraction of very cold ground as temperatures decrease rapidly in winter causes cracks that extend into permafrost. During snow melt, water fills these cracks and freezes, and over hundreds to thousands of years of repeated cracking and water infilling, large ice wedges form, represented on the ground surface as polygonal terrain. Frozen ground also limits microbial activity allowing vast stores of carbon to accumulate in permafrost, making the cryosphere an important part of the global climate system.

Interactions between the glacial and periglacial domains of the cryosphere have shaped and continue to alter landscapes around the world. The Great Lakes of North America resulted from glacial scouring, isostatic depression, and infilling from melt water as the Laurentide ice sheet retreated at the end of the Wisconsinan glaciation about 10 000 years ago. The Scandinavian fjords were similarly formed by glacial erosion. In contemporary periglacial regions, buried glacial ice in permafrost may be exposed, initiating large retrogressive-thaw slumps that dramatically alter the local landscape.

Our current understanding of the cryosphere has built upon a foundation of research spanning centuries. In addition to the varied scales of the landscape features in cold regions, the scale of our inquiry has changed over time as knowledge progressed, research interests shifted, and methods, tools, and technology improved. Modern analysis and interpretation of ice cores and geologic evidence reveals that ice ages occurred millions of years ago, while ultrasonic microphones allow sub-second measurement of frost shattering and ice-wedge cracking. These techniques were unattainable hundreds of years ago.

Cold regions — both in polar and mountain environments of the Earth — were largely ignored in early scientific thought, and were viewed with fear due to their inhospitable climates and difficult terrain. One of the earliest scientific investigations into the cryosphere dates from 1611, when Johannes Kepler wrote "Strena Seu de Nive Sexangula" while observing snowflakes from the relative safety of his home in Prague. This book examined the origin of the hexagonal snowflake, and was written as a New Year's gift to his friend.

A pioneering glaciological study occurred in 1705, when J.J. Scheuchzer hypothesized that the motion of Alpine glaciers was due to water freezing in cracks on the glacier surface, forcing the mass downslope. Subsequent research improved on this work by acknowledging gravity as the driving force of glacier flow, and the viscous nature of moving glacier ice. In the late 18th and early 19th century, the presence of erratics and moraines far away from glaciers in the Alps spurred larger-scale theorizing of past glaciations by researchers such as Louis Agassiz, which later led

Individual ice crystals of various shapes and sizes are the smallest features and building blocks of the cryosphere.

to the theories of ice ages. In contrast, and on a much smaller scale, scientists including John Tyndall, Albert Heim, Eduard Hagenbach-Bischoff, and Robert Emden refined theories on the behaviour of individual ice crystals and the plastic deformation of ice in the late 19th century (this is by no means near a complete history of the key figures in early glaciology. See Seligman's 1949 essay "Research on Glacier Flow: An Historical Outline" for an in-depth account).

Early observations of permafrost consisted of informal accounts by European explorers in sub-arctic regions of North America and Eurasia. The first scientific study of permafrost resulted from the excavation of a 116 m well into permafrost in Yakutsk, Siberia by Fedor Shergin in 1828. The excavation continued over ten years, and in 1844, Alexander Middendorff instrumented the shaft with thermometers. The ground temperature measurements provided the first evidence of the great depths to which the cryosphere extends underground. In the late 19th and early 20th centuries, investigations in Eurasia and North America examined a breadth of topics at varied spatial and temporal scales, including observations of small ice bodies, the first continental delineations of permafrost extent, and the extent of periglacial environments during the last glacial period.

From the 1930s on, cryospheric research became more established and inquiry shifted from largely vague theorizing and opportunistic sampling to more rigorous approaches. This was possible as underlying physical processes in many sub-disciplines of cold regions research were identified and better theories developed. The proliferation of statistical and computational methods after World War II enabled the scale of our inquiry to expand in both space and time. The advent of powerful computers has allowed contemporary researchers to conduct numerical simulations incorporating small-scale physical processes across broad spatial and temporal scales. For example, blowing snow models represent the behaviour of millimetre-scale snow particles over temporal scales of seconds or minutes, while an ice sheet model may calculate ice thickness on Greenland over hundreds of thousands of years. Satellite imagery and other remote sensing technologies have enabled us to observe global-scale changes in, for example, sea-ice and glacier extent, and sea-level rise — phenomena that impact humans around the world.

Our ability to study the cryosphere has never been greater, and the scale at which we choose to apply our inquiry will, as it has in the past, continue to shift. Recently, we have taken our inquisition supra caeli, searching for ice beyond our small frozen Earth. The scales of features in the cryosphere depend on physical processes controlling the configuration of ice crystals. However, the scale at which we interpret the cryosphere is wholly dependent on how we choose to discretize space and time, an entirely human pursuit.

TC Co-Editors-in-Chief

JONATHAN L. BAMBER, *University of Bristol (UK); j.bamber@bristol.ac.uk*
FLORENT DOMINÉ, *Université Laval and CNRS (Canada); florent.domine@gmail.com*
STEPHAN GRUBER, *Carleton University (Canada); stephan.gruber@carleton.ca*
G. HILMAR GUDMUNDSSON, *British Antarctic Survey, Cambridge (UK); ghg@bas.ac.uk*
MICHIEL VAN DEN BROEKE, *Utrecht University (Netherlands); m.r.vandenbroeke@uu.nl*

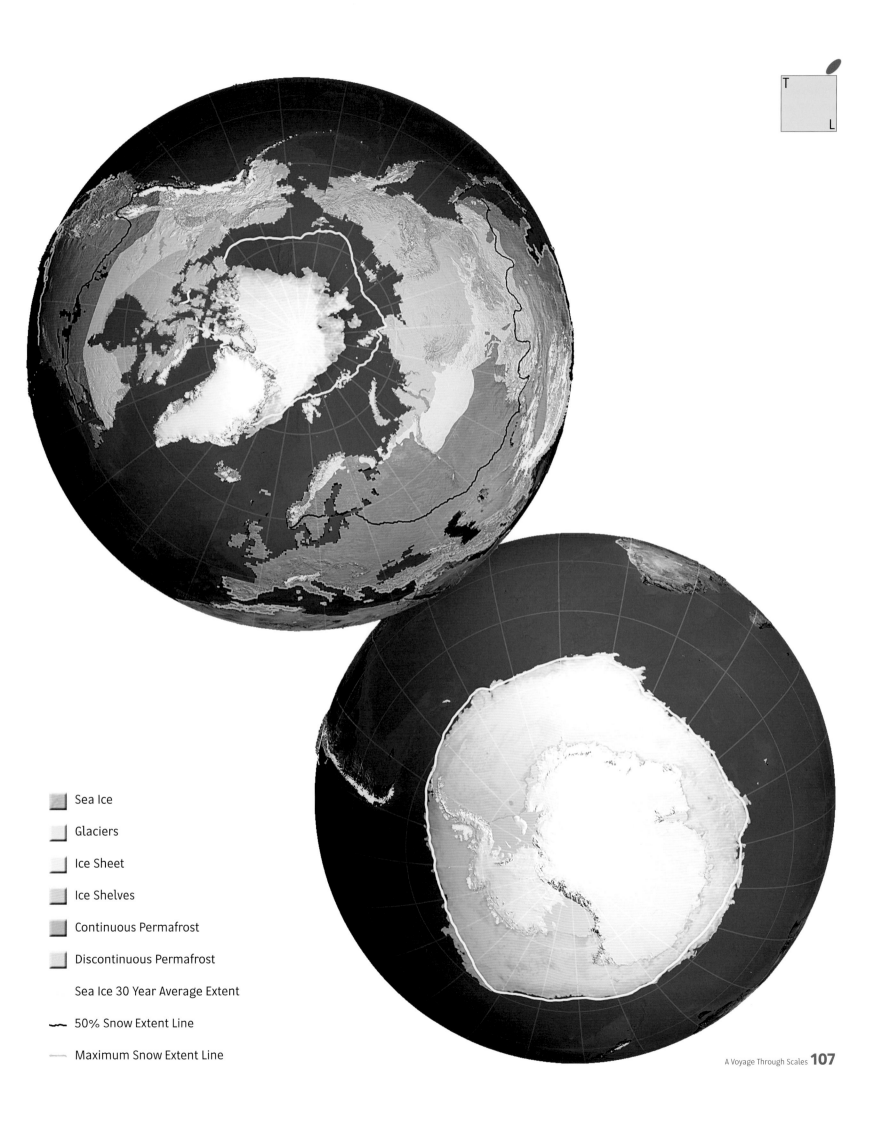

Sea Ice

Glaciers

Ice Sheet

Ice Shelves

Continuous Permafrost

Discontinuous Permafrost

Sea Ice 30 Year Average Extent

50% Snow Extent Line

Maximum Snow Extent Line

Three different features of the cryosphere: snow, glacier ice, and sea ice, covering about 48 x 42 km of Pearyland, northern Greenland.

River ice breakup http://youtu.be/0gMBQFf64JM
Largest glacier calving ever filmed http://youtu.be/hC3VTgIPoGU

Ice wedges in permafrost, represented on the ground surface as polygons,
Samoylov Island, Russia.

OPEN ACCESS — WHAT ELSE?

Martin Rasmussen
Copernicus.org

OPEN ACCESS IS FREE ACCESS. Major breakthroughs in science often arise from a close-knit network of information sharing within the research community. For centuries, publishing houses received the copyright from authors and sold their publications to fellow scientists. The European Geosciences Union and its publishing partner Copernicus Publications were among the pioneers who reversed this model some 15 years ago and introduced open-access journal publications, in which the authors pay processing charges for the editing and archiving services but retain the copyright and their articles are published online. The obvious benefits are that anyone is free to read, distribute, and adapt the work as long as the original authors are given credit — a much more powerful approach to information sharing which turns publishers from content providers to service providers. **OPEN ACCESS IS BARRIER-FREE ACCESS.** But this is just the start. There is a whole plethora of new opportunities for anyone to reuse the work. At the technical level, this has been facilitated by the EGU-Copernicus partnership through recently changing the production workflow to XML format — an extremely versatile format that is compatible with any media, from your web browser to your

iPad, e-book, and mobile phone. The meaning of the article contents is encoded rather than how it looks, which makes this concept so effective, enabling intelligent search engines and data mining — real reuse of the science rather than copy and pasting of words. Even equations can be transliterated into Braille, making this a truly barrier-free concept. **OPEN ACCESS IS A GENERAL PRINCIPLE.** Open access does even more for you. The Interactive Public Peer Review™ system of EGU-Copernicus is a very transparent way of assuring the quality of the articles, from speedy publication of the manuscripts in the journals' discussion sections, publicly accessible referee reports, and responses to the final journal publication. There is full involvement of the science community in the process through the opportunity to comment on any paper prior to publication. Open access is more than just free availability of an article: it is a commitment to the general principle of how to do community science — a general principle that will guide science communication in the 21st century, fostering creativity and excellence in research.

The 16 EGU journals on the altar of science: high quality through fast but rigorous peer review, free access and free reuse for anyone, maximum visibility and impact, community-driven — from scientists for scientists.

Open Access Peer-Reviewed Journals of EGU

http://egu.eu/publications/open-access-journals/

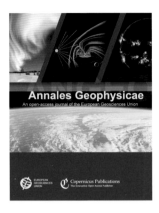

Annales Geophysicae (ANGEO)

Annales Geophysicae is an international, multi- and inter- disciplinary scientific journal for the publication of original articles and of short communications (Letters) for the sciences of the Sun-Earth system, including the science of Space Weather, the Solar-Terrestrial plasma physics, and the Earth's atmosphere.

www.annales-geophysicae.net/

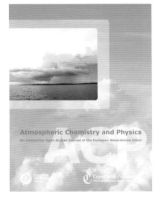

Atmospheric Chemistry and Physics (ACP) & Discussions (ACPD)

Atmospheric Chemistry and Physics is an international scientific journal dedicated to the publication and public discussion of high quality studies investigating the Earth's atmosphere and the underlying chemical and physical processes. It covers the altitude range from the land and ocean surface up to the turbopause, including the troposphere, stratosphere and mesosphere.

www.atmospheric-chemistry-and-physics.net/

Atmospheric Measurement Techniques (AMT) & Discussions (AMTD)

Atmospheric Measurement Techniques is an international scientific journal dedicated to the publication and discussion of advances in remote sensing, in-situ and laboratory measurement techniques for the constituents and properties of the Earth's atmosphere.

www.atmospheric-measurement-techniques.net/

Biogeosciences (BG) & Discussions (BGD)

Biogeosciences is an international scientific journal dedicated to the publication and discussion of research articles, short communications and review papers on all aspects of the interactions between the biological, chemical and physical processes in terrestrial or extraterrestrial life with the geosphere, hydrosphere and atmosphere. The objective of the journal is to cut across the boundaries of established sciences and achieve an interdisciplinary view of these interactions. Experimental, conceptual and modelling approaches are welcome.

www.biogeosciences.net/

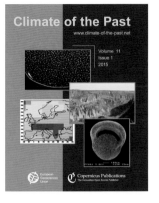

Climate of the Past (CP) & Discussions (CPD)

Climate of the Past is an international scientific journal dedicated to the publication and discussion of research articles, short communications and review papers on the climate history of the Earth. CP covers all temporal scales of climate change and variability, from geological time through to multidecadal studies of the last century. Studies focussing mainly on present and future climate are not within scope.

www.climate-of-the-past.net/

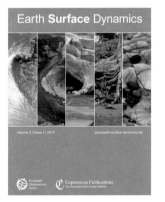

Earth Surface Dynamics (ESurf) & Discussions (ESurfD)

Earth Surface Dynamics is an international scientific journal dedicated to the publication and discussion of high quality research on the physical, chemical and biological processes shaping Earth's surface and their interactions on all scales. The main subject areas of ESurf comprise field measurements, remote sensing and experimental and numerical modelling of Earth surface processes, and their interactions with the lithosphere, biosphere, atmosphere, hydrosphere and pedosphere.

www.earth-surface-dynamics.net/

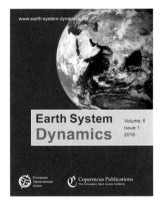

Earth System Dynamics (ESD) & Discussions (ESDD)

Earth System Dynamics is an international scientific journal dedicated to the publication of studies that take an interdisciplinary perspective of the functioning of the whole Earth system and global change. ESD publishes papers on the interactions within and between different components of the Earth system, emphasizing the underlying mechanisms from the smallest to largest scales; ways to conceptualize, model, and quantify these interactions; predictions of the overall system behavior to global changes; and impacts for Earth's habitability, humanity, and future Earth system management.

www.earth-system-dynamics.net/

Geoscientific Instrumentation, Methods and Data Systems (GI) & Discussions (GID)

Geoscientific Instrumentation, Methods and Data Systems is an open access interdisciplinary electronic journal for swift publication of original articles and short communications in the area of geoscientific instruments. A unique feature of the journal is the emphasis on synergy between science and technology that facilitates advances in GI.

www.geoscientific-instrumentation-methods-and-data-systems.net/

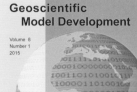

Geoscientific Model Development (GMD) & Discussions (GMDD)

Geoscientific Model Development is an international scientific journal dedicated to the publication and public discussion of the description, development and evaluation of numerical models of the Earth system and its components.

www.geoscientific-model-development.net/

Hydrology and Earth System Sciences (HESS) & Discussions (HESSD)

Hydrology and Earth System Sciences is an international two-stage open access journal for the publication of original research in hydrology, placed within a holistic Earth System Science context. HESS encourages and supports fundamental and applied research that seeks to understand the interactions between water, earth, ecosystems and man.

www.hydrology-and-earth-system-sciences.net/

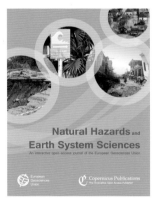

Natural Hazards and Earth System Sciences (NHESS) & Discussions (NHESSD)

Natural Hazards and Earth System Sciences is an interdisciplinary and international journal dedicated to the public discussion and open access publication of high quality studies and original research on natural hazards and their consequences. Embracing a holistic Earth System Science approach, NHESS serves a wide and diversified community of research scientists, practitioners and decision makers concerned with natural hazards detection, monitoring and modelling, vulnerability and risk assessment, and the design and implementation of mitigation and adaptation strategies, including economical, societal and educational aspects.

www.natural-hazards-and-earth-system-sciences.net/

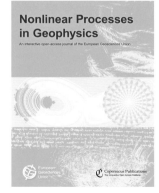

Nonlinear Processes in Geophysics (NPG) & Discussions (NPGD)

Devoted to nonlinearity research in all areas of Earth, atmospheric and planetary sciences. Nonlinear Processes in Geophysics is an international, interdisciplinary journal for the publication of original research furthering knowledge on nonlinear processes in all branches of Earth, planetary and solar system sciences. The editors encourage submissions that apply nonlinear analysis methods to both models and data.

www.nonlinear-processes-in-geophysics.net/

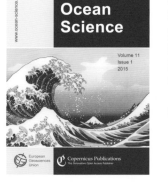

Ocean Science (OS) & Discussions (OSD)

Ocean Science is an international and free to web scientific journal dedicated to the publication and discussion of research articles, short communications and review papers on all aspects of ocean science, experimental, theoretical and laboratory. The primary objective is to publish a very high quality scientific journal with free web based access for researchers and other interested people throughout the world.

www.ocean-science.net/

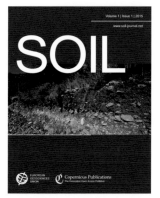

SOIL & SOIL Discussions (SOILD)

SOIL is an international scientific journal dedicated to the publication and discussion of high-quality research in the field of soil system sciences. SOIL is at the interface between the atmosphere, lithosphere, hydrosphere, and biosphere. SOIL publishes scientific research that contributes to understanding the soil system and its interaction with humans and the entire Earth system.

www.soil-journal.net/

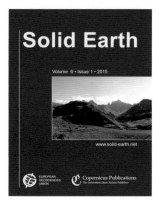

Solid Earth (SE) & Discussions (SED)

Solid Earth is an international scientific journal dedicated to the publication and discussion of multidisciplinary research on the composition, structure and dynamics of the Earth from the surface to the deep interior at all spatial and temporal scales.

www.solid-earth.net/

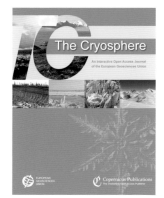

The Cryosphere (TC) & Discussions (TCD)

The Cryosphere is an international scientific journal dedicated to the publication and discussion of research articles, short communications and review papers on all aspects of frozen water and ground on Earth and on other planetary bodies.

www.the-cryosphere.net/

CREDITS

Idea Thomas Hofmann
Editors Günter Blöschl, Hans Thybo, Hubert Savenije
Concept Thomas Hofmann, Günter Blöschl, Lois Lammerhuber

Authors Günter Blöschl, Hans Thybo, Hubert Savenije, Shaun Lovejoy; Ioannis A. Daglis, Christoph Jacobi, Mike Pinnock; Laura Felgitsch, Hinrich Grothe; Andreas Richter, Thomas Wagner; Michael Bahn, Katja Fennel, Jürgen Kesselmeier, Wajih Naqvi, Albrecht Neftel; Tom Coulthard, Jérôme Gaillardet, Frédéric Herman, Niels Hovius, Douglas Jerolmack, Andreas Lang; Somnath Baidya Roy, Axel Kleidon, Valerio Lucarini, Ning Zeng; Jothiram Vivekanandan, Ari-Matti Harri, Håkan Svedhem; Alberto Guadagnini, Erwin Zehe, Hubert H.G. Savenije, Alison D. Reeves; Bruce D. Malamud; Ana María Mancho, Jezabel Curbelo, Stephen Wiggins, Víctor José Garcia-Garrido, Carolina Mendoza; John M. Huthnance; John N. Quinton, Jorge Mataix-Solera, Eric C. Brevik, Artemi Cerdá, Lily Pereg, Johan Six, Kristof van Oost; Charlotte M. Krawczyk; Brendan O'Neill; Martin Rasmussen

Photography and art director Lois Lammerhuber
Graphic design Martin Ackerl, Lois Lammerhuber
Typeface Lammerhuber by Titus Nemeth
Digital post production Birgit Hofbauer
Project coordination Thomas Hofmann, Johanna Reithmayer
Proofreading Gemma Carr

Cover: See credits page 86; Endpaper 1: Boris Murdja; Endpaper 2: Department für Lithosphärenforschung, Universität Wien, Christian Köberl; 2: Francis Giacobetti; 4: NASA Earth Observatory; 6: Yann Arthus-Bertrand/Altitude; 8: Dennis Kunkel Microscopy, Inc./Visuals Unlimited, Inc.; 10: Dr. Eric Condliffe/ Visuals Unlimited, Inc.; 12: Peter Ginter; 18: Adapted from Lovejoy and Schertzer (2013): The Weather and Climate: Emergent Laws and Multifractal Cascades, Cambridge Univ. Press; 20: Bjorn Jorgensen/www.arcticphoto.no; 22: From the book Human Footprint, eoVision, www.eovision.at; 24: Wim van Egmond/Visuals Unlimited, Inc.; 26: N. Alzate, Univ. of Aberystwyth, Wales. The algorithm used for deriving this image: "Multi-scale Gaussian Normalization for Solar Image Processing" by Huw Morgan and Miloslav Druckmuller (ApJ 2014); 28: NASA; 29: ESA/Cluster; 30: NASA/SOHO; 31: Olivier Grunewald; Walter Myers; NASA; 32: Michael Gauss; 34: Max-Planck-Institut für Chemie; Max-Planck-Institut für Chemie; Copyright: Laura Felgitsch, Vienna University of Technology, Institute of Materials Chemistry; 35: Karlsruhe Institute of Technology (KIT); 36: NASA; 37: Karlsruhe Institute of Technology (KIT); Dartmouth Electron Microscope Facility/Dartmouth College; 38: NOAA; 41: Nadeau, P., and Dalton, M., 2009, Report on UV camera field campaign, Fuego and Santiaguito volcanoes, Guatemala, December 2008-January 2009, unpublished informal report accessed January 2010; 42: NASA; 43: NASA; 44: NASA; 47: Raphael Felber/Agroscope ISS-Switzerland; 48: James Nicholson, NOAA CDHC; 49: David Doubilet/National Geographic Creative; 50: UDINE20.it; 52: From Braun et al. (2015): Earth Surf. Dynam., 3, 1-14; 53: ESA/DLR/FU Berlin (G. Neukum); 54: iStockfoto/ericfoltz; 56: Evan M. Butterfield; 59: NASA/GSFC/MITI/ERSDAC/JAROS, and U.S./Japan ASTER Science Team; 60: Jörg Weiß, Mikroskopisches Kollegium Bonn, www.mikroskopie-bonn.de; 61: NASA; 62: David Roberts; 65: UCAR, photo by Michael Dixon, NCAR; 68: Merlin UK/International Federation of Red Cross and Red Crescent Societies; 70: Garing et al. (2015) doi:10.1007/s11242-015-0456-2; 72: Ilan Shacham/Getty; 74: Gene Blevins/Reuters/Corbis; 77: Peter Thompson; 78: NASA (Visible Earth), USA National Hurricane Center, image by Wikipedia users Nilfanion and Supportstorm; 80, 82, 83: Jezabel Curbelo and Ana María Mancho; 84: NASA/MODIS; 86: Snapshots of the surface current speed as simulated by a hindcast run of a high resolution ocean model forced with past atmospheric conditions. Courtesy of Dr A. C. Coward, Marine Systems Modelling group, National Oceanography Centre, UK (www.noc.ac.uk). These simulations are funded by the Natural Environment Research Council and performed on the ARCHER UK National Supercomputing Service (www.archer.ac.uk); 89: Jean Guichard; 90: David Doubilet/Undersea Images Inc.; 91: Michael Catania/Solent News & Photo Agency; 92: Jim Richardson; 95: Jorge Mataix-Solera; 96: Photo courtesy of Tim McCabe, USDA-NRCS; Jorge Mataix-Solera; 97: Jorge Mataix-Solera; The University of Jordan/Wadah Mahmoud; 98: Integrated Ocean Drilling Program/John Beck; 101: p. 253 from: Stüwe und Homberger (2011): Die Geologie der Alpen aus der Luft. — Weishaupt; 103: Lucas Lourens; 104: Don Komarechka; 107: NASA/Goddard Space Flight Center Scientific Visualization Studio; 108: NASA/GSFC/METI/Japan Space Systems and U.S./Japan ASTER Science Team;109: Sebastian Zubrzycki (distributed via imaggeo. egu.eu); 110: NASA/Bill Putman; Endpaper 3: David Nunuk/Science Photo Library/Corbis; Back Cover: NASA/SDO/AIA. *Every care has been taken to identify and acknowledge the owners of rights in the content of the work. Should concerns nevertheless arise, the publisher will welcome your comments.*

Printing Bösmüller, Stockerau, Austria
Binding Kösel, Altusried, Germany
Paper GardaPat 11, 170 g/m²

Responsible at EUROPEAN GEOSCIENCES UNION Günter Blöschl
EUROPEAN GEOSCIENCES UNION Luisenstraße 37, 80333 Munich, Germany
www.egu.eu

Managing director EDITION LAMMERHUBER Silvia Lammerhuber
EDITION LAMMERHUBER Dumbagasse 9, 2500 Baden, Austria
edition.lammerhuber.at

Günter Blöschl is Professor of Hydrology, Director of the Centre for Water Resource Systems, and Head of the Institute of Hydraulic Engineering and Water Resources Management at the Vienna University of Technology. He has published extensively on subjects related to hydrology and water resources. He is Fellow of the American Geophysical Union, member of the German Academy of Science and Engineering, and recipient of an Advanced Grant of the European Research Council (ERC). He is the President of the European Geosciences Union.

Hans Thybo is Professor of Geophysics and Head of the Geophysics Laboratory at University of Copenhagen. His extensive publications cover subjects related to plate tectonics, lithosphere evolution, seismology and application of geophysics to exploration for resources. He is Vice-president of the Royal Danish Academy of Science and Letters and member of other academies, including Academia Europae. He is Fellow of Geological Society of America and Royal Astronomical Society, London. He is the incoming President of the European Geosciences Union.

Hubert Savenije is Professor of Hydrology and Head of the Water Resources Section at Delft University of Technology. He is specialized in the Hydrology of Catchments, River Basins and Estuaries. He received the Henry Darcy medal and Alexander Von Humboldt medal of the European Geosciences Union and is Fellow of the American Geophysical Union. He is the Chair of the Publications Committee of EGU.

A VOYAGE THROUGH SCALES is the theme of the
2015 General Assembly of the European Geosciences Union.

Milky Way and observatories on the summit of Mauna Kea, Hawaii. Subaru Telescope, Keck 1 and 2 telescopes, NASA Infrared Telescope Facility (from left to right).